INSTANT
SCIENCE

Portable Press
An imprint of Printers Row Publishing Group
10350 Barnes Canyon Road, Suite 100, San Diego, CA 92121
www.portablepress.com • mail@portablepress.com

Copyright © 2019 Welbeck Non-fiction Limited

All rights reserved. No part of this publication may be reproduced, distributed, or transmitted in any form or by any means, including photocopying, recording, or other electronic or mechanical methods, without the prior written permission of the publisher, except in the case of brief quotations embodied in critical reviews and certain other noncommercial uses permitted by copyright law.

Printers Row Publishing Group is a division of Readerlink Distribution Services, LLC.
Portable Press is a registered trademark of Readerlink Distribution Services, LLC.

Correspondence regarding the content of this book should be addressed to Portable Press, Editorial Department, at the above address. Author and illustration inquiries should be addressed to Welbeck Publishing Group, 20–22 Mortimer St, London W1T 3JW.

Portable Press
Publisher: Peter Norton
Associate Publisher: Ana Parker
Editor: April Graham Farr
Senior Product Manager: Kathryn C. Dalby
Produced by Welbeck Non-fiction Limited

Library of Congress Control Number: 2019945703

ISBN: 978-1-64517-056-3

Printed in Dubai

23 22 21 20 19   1 2 3 4 5

All illustrations provided by Noun Project with the exception of pg31 Oleg Alexandrov via Wikimedia Commons, pg31 Tttrung via Wikimedia Commons, pg33 Alevtina Vyacheslav/Shutterstock, pg72 bhjary/Shutterstock, pg81 Trekandshoot/Shutterstock, pg82 Mars Brashok/Shutterstock, pg87 Julie Deshaies/Shutterstock, pg90 Magnetix/Evannovostro/Shutterstock, pg103 ShadeDesign/Shutterstock, pg109 Sakurra/Shutterstock, pg110 Aldona Griskeviciene, pg117 Designa/Shutterstock, pg121 Aldona Griskeviciene/Shutterstock, pg 125 Brenik/Shutterstock, pg136 Vecton/Shutterstock, pg157 Nanmulti/Shutterstock, pg166 Soleil Nordic/Shutterstock

INSTANT SCIENCE

KEY THINKERS, THEORIES, DISCOVERIES, AND INVENTIONS EXPLAINED ON A SINGLE PAGE

JENNIFER CROUCH

PORTABLE PRESS

San Diego, California

CONTENTS

8   Introduction

MATHEMATICS

10   Numbers
11   Counting Systems
12   Symmetry
13   Euclid's Elements
14   Tessellation
15   Platonic Solids
16   Al-Khwarizmi
17   The Fibonacci Sequence
18   Infinity
19   Pi
20   Prime Numbers
21   Calculus
22   Logic
23   Logarithms
24   Probability and Statistics
25   Chaos
26   Imaginary Numbers
27   Non-Euclidean Geometry
28   Fermat's Last Theorem
29   Euler's Number
30   The Mandelbrot Set
31   Topology

PHYSICS

32   Planetary Resonance
33   Al-Battani
34   Simple Harmonic Motion
35   Optics
36   Sound and Acoustics
37   Telescopes
38   Work Done, Power, and Energy
39   Kepler's Laws
40   Noether's Conservation Laws
41   Newton's Equations of Motion
42   Gravitational Constant
43   Faraday and Electromagnetism
44   Thermodynamics
45   Absolute Zero
46   Maxwell's Equations
47   Maxwell–Boltzmann Distribution
48   Discovery of the Electron
49   Young's Double-Slit Experiment

50	The Photon	70	Stars, the Sun, and Radioactivity
51	The Rutherford Atom	71	Solar Systems
52	Marie Curie and Radioactivity	72	Space Observatories
53	The Photoelectric Effect	73	Galaxies
54	General and Special Relativity	74	Spectometry
55	Schrödinger and the Wave Equation	75	Exoplanets
56	The Uncertainty Principle	76	Meteors, Asteroids, and Comets
57	Enrico Fermi and Beta Decay	77	Pulsars and Jocelyn Bell Burnell
58	Electron States, Quantum Numbers	78	Measuring the Universe
59	Dirac and Antimatter	79	Black Holes
60	Feynman Diagrams	80	Time Dilation
61	The Manhattan Project	81	Cosmic Microwave Background Radiation
62	The Standard Model	82	Cosmic Foam
63	The Wu Experiment	83	The Big Bang
64	Neutrino Oscillation	84	Charge Parity Violation
65	The Higgs Boson	85	Dark Energy and Dark Matter
66	Quantum Electro Dynamics (QED)	86	Mysteries (the Multiverse, Super Symmetry, and String Theories)
67	Quantum Chromodynamics (QCD)		
68	Nuclear Fission Reactor		
69	Particle Accelerator		

CHEMISTRY

87	The Periodic Table	95	Hydrogen Bonding and Water
88	Carbon Dating	96	States of Matter
89	Intramolecular Bonding	97	Chirality
90	Intermolecular Bonding	98	Macromolecules
91	Chemical Reactions	99	Polymers
92	Organic Chemistry	100	Hydrophilic and Hydrophobic
93	Inorganic Chemistry	101	Protein Crystallography
94	Power of Hydrogen		

BIOLOGY & MEDICINE

102	DNA and Photo 51	117	Symbiosis
103	The Central Dogma of Biology	118	Microbiome
104	Cells	119	Evolution
105	The Microscope	120	Genetics and Mutation
106	Microbiology	121	Zoology
107	Pasteurization	122	Reproduction and Cloning
108	Vaccination	123	Stem Cells
109	Bacteriology	124	Systems of the Body
110	Virology	125	Human Anatomy
111	Extremophiles	126	Immunology
112	Biomaterials	127	Circulation of the Blood
113	Fungi	128	Parasitology
114	The Discovery of Penicillin	129	Neuroscience
115	Photosynthesis	130	Surgery
116	Multicellularity		

GEOLOGY & ECOLOGY

- 131 Life History
- 132 Principles of Ecology
- 133 Trophic Cascades
- 134 The Earth's Oceans
- 135 Extinctions
- 136 Diversity and Population
- 137 Plate Tectonics
- 138 Atmospheric Physics
- 139 Biogeochemical Cycles
- 140 Hydrological Cycle
- 141 Carbon Cycle
- 142 Rock Cycle
- 143 Geomagnetism
- 144 Bioaccumulation
- 145 Human Made Climate Change

TECHNOLOGY

- 146 The Library of Alexandria
- 147 The Circumference of the Earth
- 148 The Measurement of Time
- 149 Ismail Al Jazari
- 150 Movable Type
- 151 Construction
- 152 Heat Engine
- 153 Energy Storage
- 154 The Computer
- 155 Electronics
- 156 Alan Turing
- 157 Photography
- 158 Radar and Sonar
- 159 Information
- 160 GPS
- 161 Space Travel
- 162 Programming
- 163 Buckminster Fuller
- 164 Magnetic Resonance Imaging
- 165 The Internet
- 166 Genetic Engineering
- 167 3-D Printing
- 168 Touch Screens
- 169 Algorithms and AI

- 170 Inspiring Scientists
- 171 Data Sheets
- 174 Glossary
- 175 Further Reading

INTRODUCTION

Science describes the behavior of natural phenomena, such as electromagnetism and gravity, and uses the language of mathematics to articulate these descriptions—but science is not nature itself. As physicist Niels Bohr once put it: physics is not nature; it's what we can say about nature.

Science (like all knowledge) is collective and formed through collaborative human behaviors that involve methods of performing, developing, and administering scientific experiments; analyzing data; improving the accuracy of tools; and understanding measurements. Science can predict and explain the behavior of matter and is essential in the development of technology, which can be used for good or to cause harm. It is also dependent on public funding, subject to the whims of political interest and economic investment, and thus it is not neutral and does not escape society, social bias, resource-related issues, or politics.

This book presents knowledge circa 2020 and perhaps in five, ten, fifty, or one hundred years the contents of this book will need to be completely revised as new discoveries are made. The aim of this book is to summarize 160 different topics, providing short descriptions, brief histories, and explanations of scientific ideas across math, physics, chemistry, biology, medicine, ecology, geology, and technology. As each topic is limited to one page, make use of the glossary where necessary; use the units table, scientists, data sheets, and equations at the back.

Scientific knowledge and practice evolve with technology and the ever-changing design of experiments, as well as with new ways of communicating and collaborating. The scientific method formulates hypotheses that are developed into theories. A theory in the context of science is very different from the everyday use of the word theory: a scientific theory has a lot of evidence supporting it—i.e., the theory of electromagnetic induction, or relativity, or evolution.

Theories become accepted through repeatable observations drawn from experimental and measurement-based testing, which different groups of scientists seek to falsify (prove false) in order to avoid biases. The more we discover and the more precise technology gets, the more our theories can be challenged. The theories that are extensively challenged and still hold true become accepted.

Science is collaborative and dependent on the technical, intellectual, and emotional efforts of large groups of people working together on international projects that contribute toward generating our shared scientific knowledge. Science is a culture within cultures and depends on interactions among people working in teams in very specific environments (i.e., laboratories) that require the use of specialist equipment that needs to be designed and made from materials extracted from the Earth—which, as we will discover, has its own problems. Once laboratories are built they must then be managed.

Science is difficult to do. In practice, it is full of tricky technological challenges, experiments failing, cell cultures dying, magnet cooling systems exploding, and things generally going wrong. It's complex, fiddly, and confusing at the best times. In fact, much of the history of science is buried away in its failures but these (while not documented in this book) are important to acknowledge, rather than perceive of science as a series of seamless, glorious, paradigm-shifting innovations and discoveries. Much toil goes into creating our descriptions of nature.

In addition to this, the personal challenges encountered by individuals and groups who have been discriminated against and experienced racism, homophobia, sexism, classism, and bullying in these social, albeit scientific, spaces must continue to be confronted. As the CERN's Particles for Justice website states:

"The humanity of any person, regardless of ascribed identities such as race, ethnicity, gender identity, religion, disability, gender presentation, or sexual identity is not up for debate." (source: particlesforjustice.org)

Science is powerful and affects us all, from the internet, to the weaponization of science in warfare, to green energy, architecture, engineering, and developments in medicine and surgery. We all have a stake in it and a right to understand the wonderful things it tells us about our universe so that we can think about and question who it benefits, who it harms, and how it is being used here on Earth.

NUMBERS

Understanding quantities and concepts of numbers is something all humans learn and need to do.

NUMBERS THROUGHOUT HISTORY

With gradual changes in lifestyle, traveling, growing crops, keeping track of resources, trading, and cultural exchange, new ways of **writing numbers** emerged.

THE ISHAGO BONE
The **Ishango Bone** is an ancient African artifact that is 20,000 years old. It is the earliest example of representing numbers as a series of **tally lines**.

SUMERIAN
Sumerian artifacts from at least 4000 BC demonstrate the **earliest evidence of arithmetic** being used. Sumerian counting followed a **sexagesimal (base 60) counting system**.

ANCIENT EGYPTIAN
The Ancient Egyptians used a **hieroglyphic number system** to express geometric concepts and plan the building of the Pyramids.

MAYAN
The Ancient **Mayan** system was a **base 20 system**.

ROMAN NUMERALS
Roman numerals were inherited from the Etruscan civilization of Tuscany.

I	II	III	IV	V
1	2	3	4	5
VI	VII	VIII	IX	X
6	7	8	9	10

L	C	D	M
50	100	500	1,000

MODERN NUMBERS
The number systems commonly used today exist thanks to **Arabic and Hindu mathematicians**. Two ancient Indian mathematicians are credited with sharing this system with the world: **Brahmagupta** from the sixth century BC and **Aryabhata** from the fifth century BC.

CHINESE NUMBERS
Although Hindu–Arabic numerals are used in **China**, two other numeral systems are also employed there: one in which characters spell out the number for everyday writing (**simple numerals**); and one for financial use (**complex numerals**), historically used to protect against fraud.

Even nonhuman creatures depend on an internal concept of "**amount**." Examples include **small fish in large shoals, murmurations of starlings**, and **many types of collective animal behavior**. Bees even count landmarks between their hives and food sources.

COUNTING SYSTEMS

Different counting systems are used for different purposes.

DECIMAL

The **decimal system** is the most common counting system. It operates to **base 10**, meaning that counting happens in units of tens. **The decimal system uses place values and decimal points, and the number "0" to fill empty spaces.**

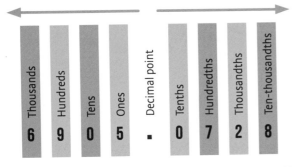

8-BIT: INTERESTING BINARY NUMBERS

Eight = 1000; 2 x 2 x 2 (three 0s)
Sixteen = 10000; 2 x 2 x 2 x 2 (four 0s)
Thirty-two = 100000; 2 x 2 x 2 x 2 x 2 (five 0s)
8-bit integers are used in programming. Comparing an 8-bit image to a 64-bit image tells us that we can use the 8-bit integer system to increase resolution in computer graphics.

BINARY

Binary is a **base 2** counting system that **uses only 0 and 1 to represent all the numbers in the decimal system.**

COUNTING IN BINARY

1. To write numbers in binary, add a "1" to the front of the number and "reset" digits to 0.
2. Fill the placeholder "0" with "1" from right to left.
3. Add a "0" on the end of the next number once you have filled your binary number with 1s.

- **0** = zero
- **1** = one
- **10** = two
- **11** = three
- **100** = four
- **101** = five
- **110** = six
- **111** = seven
- **1000** = eight (notice that at 8 the 0's reset)
- **1001** = nine
- **1010** = ten

HEXADECIMAL

Hexadecimal (or hex) is a **base 16 system** (think doubling 8-bits!) used to simplify binary numbers.

SEXAGESIMAL

This is **base 60**, and was first used by the ancient **Sumerians** in the third millennium BC. The ancient **Babylonians** used it, too. **It is still used today to measure seconds, hours, angles, and geographic coordinates.**

BASE 12

A **base 12 system** (i.e., inches, feet, and hours per day) uses base 12 counting. A 24-hour clock is simply double the 12-hour clock and doesn't need the a.m. or p.m. signs.

SYMMETRY

Mathematical symmetries can be observed as a spatial relationship; or through geometric transformations, rotation, or scaling, and even in time.

SYMMETRY RULES

If an object can be divided into two or more identical halves across a plane, or can be scaled, rotated, or reflected and still look the same, then that object has **geometric symmetrical properties**.

TYPES OF SYMMETRY TRANSFORMATION

Reflectional symmetry is **mirror symmetry** or **bilateral symmetry**, where a line passes through an object and identical or mirror-image halves can be seen on either side of the plane.

Rotational symmetry occurs when a shape is rotated about a fixed point without changing the overall shape.

Radial symmetry is rotated around a central axis, such as in starfish, jellyfish, and anemones. A **mandala** is also designed to have radial symmetry.

Translational symmetry means a shape can be translated (moved across a surface) without changing its overall shape.

Helical symmetry is a combination of translated and rotated symmetry, expanding in **three-dimensional space**. The path (or line) on which it grows is known as the **"screw axis."**

Scale symmetry occurs when a shape is **expanded** or **contracted**. A well-known example of scale symmetry is seen in **fractals**. Progressive use of scale symmetry using an equilateral triangle in the **"Koch snowflake"** can be seen above.

Glide reflection symmetry and **rotoreflection symmetry**.

SNOWFLAKES

Snowflakes are never completely symmetrical but, in a controlled environment, **they reliably exhibit six-fold radial symmetry**. The crystalline structure of ice is hexagonal and a tiny hexagonal crystal forms where this first starts to grow, called the point of **nucleation**.

CUBES

Of course, shapes can also be symmetrical in three dimensions; for instance, **a cube has nine symmetry planes**.

EUCLID'S ELEMENTS

Euclid's Elements is a collection of axioms (logical arguments) describing geometric relationships on a two-dimensional plane. Euclidean geometry does not work on curved surfaces.

EUCLID'S POSTULATES:

1. A straight-line segment **can be drawn joining any two points**.

2. Any straight-line segment **can be extended indefinitely in a straight line**.

3. Given any straight-line segment, **a circle can be drawn having the segment as radius and one endpoint as center**.

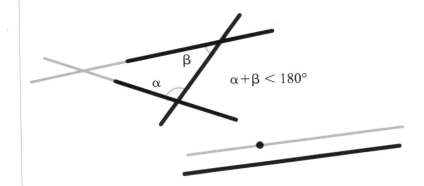

4. **All right angles are equal (congruent).**

5. If a straight line crossing two straight lines **creates interior angles on the same side that are smaller than two right angles**, then **the two straight lines extended indefinitely would eventually converge and cross each other on the same side of the line as the two interior angles**.

Euclid's fifth postulate cannot actually be proven as a theorem, though the feat has been attempted by many people. Euclid published many more postulates; these are just the first five.

MATHEMATICS

TESSELLATION

Tessellation is the arrangement of shapes fitted together in a repeated pattern, with no gaps or overlaps.

REGULAR TESSELLATION

A pattern made using the same shape with no gaps or overlaps. **Hexagons**, regular **tetrahedrons**, and **triangles** can **regularly tessellate**.

SEMI-REGULAR TESSELLATION

A pattern made by combining different polygons with no gaps or overlaps. **Hexagons** and **triangles** can still be used but also **pentagons**, **heptagons**, and **octagons**.

INTERIOR ANGLES

To explain why there are only certain shapes that can be regularly tessellated, we think interior angles. On a 2-D surface there are **360° around a circle**. For there to be no gaps or overlaps, the internal angles of a polygon being fanned out around a point must add up to 360°.

PENTAGONS

It's not possible to regularly tile a pentagon. Its interior angles do not add up to 360°. The interior angles of a pentagon measure 108°.

TILING AROUND A POINT EITHER:
- leaves a gap: 3 × 108 = 324 (less than 360°), or
- creates an overlap: 4 × 108 = 432 (greater than 360°)

But it is possible to create some very interesting semi-regular tiling patterns with the space left over.

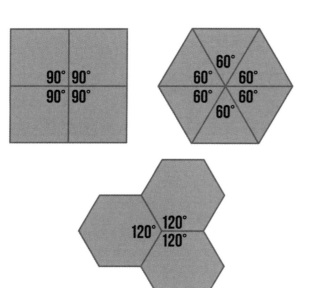

IRREGULAR TILING

This makes use **of irregular shapes**. Anything goes with **irregular tiling**! You just need to fill the space with random polygons.

PENROSE TILING

Penrose tiles are pairs of shapes that tile the plane **aperiodically** (without repeating), using a **"kite"** and **"dart"** shape. Tiles must be arranged so that the kite and dart never form a **rhombus**.

THE ALHAMBRA

The **Alhambra palace and fortress in Granada, Spain**, was built by the Moors in AD 889. It contains intricate and beautiful examples of 2-D tiling throughout.

PLATONIC SOLIDS

Platonic solids are 3-D shapes. Each face of these shapes is a regular polygon where each polygon meets at a vertex.

REGULAR POLYGONS

These have sides of the same length and internal angles of the same side, e.g., an **equilateral triangle**, **square**, or **regular pentagon**.

THE PLATONIC SOLIDS

For a 3-D shape to be a **Platonic solid**, all the faces must be the same and all faces must be regular polygons. There are only five Platonic solids.

PLATO'S WORLD OF THE FORMS

Ancient philosopher **Plato** thought the above shapes were magical. He also believed that there was a **"world of the forms"**—a place, or superpositioned space, where the Platonic shapes actually existed. In **Platonic Mysticism**, these shapes were also believed to embody the supposed **five elements of Earth, Fire, the Cosmos, Water, and Air**.

ABSTRACTIONS AND REALITY

An ongoing discussion in science, mathematics, and philosophy is about **whether abstractions exist in the world or in the mind**. Is the way we understand the universe that we observe dependent on the abstractions we use to describe it? **Are mathematical concepts invented or are they discovered?**

OTHER 3-D SHAPES

There are many more shapes that are possible to discover/invent if we use other kinds of polygons or combinations of different polygons.

STELLATED DODECAHEDRON
The **stellated dodecahedron** is made from **isosceles triangles**.

PENTAKIS DODECAHEDRON
Also called a **kisdodecahedron**, this is a dodecahedron with a **pentagonal pyramid** covering each face.

ARCHIMEDEAN SOLIDS

The **Archimedean solids** are made from **two or more different types of polygon arranged around each vertex with all sides the same length**. There are thirteen Archimedean solids.

Stellated dodecahedron

Pentakis dodecahedron

AL-KHWARIZMI

Al-Khwarizmi lived from AD 780 to 850 and was a Persian mathematician born in Uzbekistan. He was a teacher at the Baghdad House of Wisdom and is known as the inventor of algebra.

Between 813 and 833, al-Khwarizmi published his algebraic treatise in a book called ***The Compendious Book on Calculation by Completion and Balancing***—a **codification of Babylonian, Hindu, and Islamic numerical and mathematical knowledge**, and one of the earliest demonstrations of **linear and quadratic equations**.

- *Al-jabr* means "restoring" or "completion" and includes **removing negative numbers, roots, and squares** to simplify an equation.
- *Al-muqabala* or **"balancing"** is still used in solving equations.

THE FIRST TO COMPLETE THE SQUARE

Completing the square is a technique for **converting a quadratic polynomial with unknown properties** that one wants to know into a form where it is easier to work out these unknowns. For example, from $ax^2 + bx + c$ to the form $a(x - h)^2 + k$ (where h and k have a value).

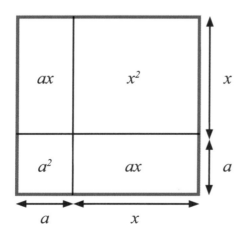

MODERN PROCESS FOR FACTORIZING QUADRATIC EQUATIONS

1. Rearrange the equation, making it equal to zero.
2. Factorize, e.g., rearrange $2x^2 + x - 3$ into the form $(2x + 3)(x - 1)$.
3. Set each factor equal to zero.
4. Solve each of these equations.
5. Verify your solution by **substitution**.

THE QUADRATIC FORMULA

$$x = \frac{-b \pm \sqrt{b^2 - 4ac}}{2a}$$

COMPLETING THE SQUARE EXAMPLE:

$$x^2 - 10x + 25 = -16 + 25$$
$$(x - 5)^2 = 9$$
$$x - 5 = \pm\sqrt{9}$$
$$x - 5 = \pm 3$$
$$x = 5 \pm 3$$
$$x = 8 \text{ or } x = 2$$

THE FIBONACCI SEQUENCE

The Fibonacci sequence is an infinite sequence that starts with 0 and 1. The Fibonacci formula is "to create the next number, add the two preceding numbers." Italian mathematician Leonardo Fibonacci, born around AD 1170, wrote a book that popularized the ratio, but many Egyptians and Babylonians before him had explored its existence.

It can be defined by the following **linear recurrence equation**: $F_n = F_{n-1} + F_{n-2}$

If the first term is $F_0 = 0$

and $n = 1$

the sequence follows:

0, 1, 1, 2, 3, 5, 8, 13, 21...

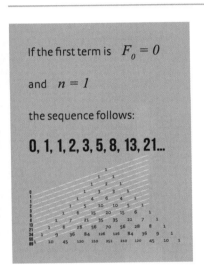

THE GOLDEN RATIO

The golden ratio is a **proportion** described by the following ratio: **$a + b : a$**

$a + b$ is to a as a is to b

The golden ratio is an **irrational number** that is calculated using the following formula:

$a \div b = (a + b) \div a = 1.6180339887498948420...$

FIBONACCI AND THE GOLDEN RATIO

By dividing each Fibonacci number by the previous number, the values approach the golden ratio. In math speak, **they are said to converge toward the golden ratio.**

$1 \div 1 = 1$
$2 \div 1 = 2$
$3 \div 2 = 1.5$

... and so on up to

$144 \div 89 = 1.6179$

THE GOLDEN SPIRAL

This spiraling shape is drawn by using lines whose length difference increases by proportion of the golden ratio; **a continuous arch can bridge the lines, resulting in a spiral.**

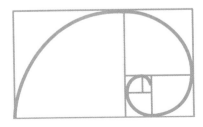

HUMANS LOVE PATTERNS

We are a pattern-seeking (and pattern-loving) species, so all these numbers seem almost magical. But thinking of them as magical generates misinformation about the golden ratio. **Many claims concerning the ubiquitous cosmic power of the Fibonacci sequence and golden ratio are measurably false.**

PLANT GROWTH

The golden ratio captures some types of plant and animal growth, but beware—**there are just as many plants and animals that don't follow this rule or present such geometric proportions**. Pineapples, sunflowers, and pinecones display spacing of seeds and leaves that matches the golden ratio.

INFINITY

The concept of infinity is used by mathematicians and physicists to describe a quantity without end.

The symbol for infinity is

DIFFERENT KINDS OF INFINITY AND SET THEORY

Infinity is more than just a really large number. There are different types of infinity. **Set theory is an area of mathematics that investigates categories.** It seeks to specify the parameters that define different sets of numbers, exploring and explaining their properties. **Infinities, like other mathematical numerical concepts, can be organized using sets.**

SETS OF INFINITIES:
- ∞ Positive numbers
- ∞ Negative numbers
- ∞ Fractions
- ∞ Irrational numbers
- ∞ Square numbers

INFINITE DECIMAL PLACES

The infinitesimal is also a type of infinity, and we can think of an infinite number of decimal places between every number. Some fractions also give infinite decimal places, such as 1/3.

$$1/3 = 0.33333\ldots$$

If a number has terminating or repeating decimals, and can be represented as a fraction, then it is a rational number. **Numbers that cannot be written as a ratio of two integers are called irrational numbers.** For more on this, see the topic Imaginary Numbers (page 26).

DIVIDE BY 0

When we divide any number by **0, the answer we get is not infinite; it is undefined.** That is, it can't be placed on the number line.

ASYMPTOTIC ANALYSIS

The functions below are asymptotic. **An asymptotic curve forever approaches a value but never reaches it.**

The functions $y = tan(x)$ and $y = 1/x$ are said to be asymptotically discontinuous

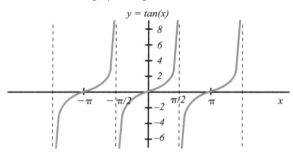

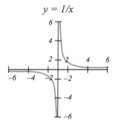

DISCONTINUOUS AND CONTINUOUS FUNCTIONS

A discontinuous function is not a continuous curve but looks like separate curves. To draw a discontinuous function, you need to lift your pencil up at least once to draw the curve. A continuous function flows and connects across the plane without breaks.

PI

*The number π (also called Archimedes' constant) is a mathematical constant.
It is defined as the ratio of a circle's circumference to its diameter.*

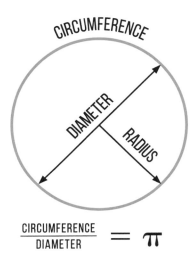

$$\frac{\text{CIRCUMFERENCE}}{\text{DIAMETER}} = \pi$$

π is an irrational number and has infinite decimal places

3.14159265359...

TRANSCENDENTAL NUMBER

π is **a transcendental number, not an algebraic number**.

ALGEBRAIC NUMBERS

All algebraic numbers such as whole numbers and fractions can be expressed as **the square root of a non-zero "polynomial" equation with whole number, fraction, or other integer coefficients**.

For example, a quadratic equation is a type of polynomial equation that has this form:

$x^2 + bx + c = 0$

Ordinary algebraic numbers **can be expressed using polynomial equations (such as a quadratic)** where b and c are whole numbers or fractions.

What makes π special is that it cannot be expressed using these terms. This is why it is impossible to square the circle.

Area of a circle = πr². A circle of radius 1 has area π. A square with area equal to π must have a length square root of π. **But it is impossible to express a transcendental number such as π as a polynomial equation. This is why it is impossible to square the circle.**

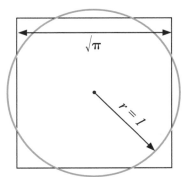

POLYGON APPROXIMATION

π can be approximated by using polygons of increasing numbers of sides. **This method will never get you exactly to π but you can asymptotically approach it with a polygon of ever increasing sides.**

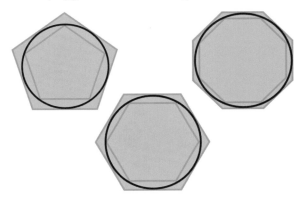

PRIME NUMBERS

A prime number is a number that is only divisible by one and itself. An infinite number of primes are thought to exist, but they are fewer and further between as we move away from zero and toward infinity.

The first few **prime numbers** are: 2, 3, 5, 7, 11, 13, 17, 19, 23, 29, 31, 37, 41, 43, 47, 53, 59, 61, 67, 71, 73, 79, 83, 89, 97, 101...

They can be arranged in a grid to help us find patterns.

1	2	3	4	5	6	7	8	9	10
11	12	13	14	15	16	17	18	19	20
21	22	23	24	25	26	27	28	29	30
31	32	33	34	35	36	37	38	39	40
41	42	43	44	45	46	47	48	49	50
51	52	53	54	55	56	57	58	59	60
61	62	63	64	65	66	67	68	69	70
71	72	73	74	75	76	77	78	79	80
81	82	83	84	85	86	87	88	89	90
91	92	93	94	94	96	97	98	99	100

EUCLID'S PROOF

Simply states that there exists **an infinite number of primes**.

THE PRIME NUMBER THEOREM (PNT)

PNT tells us **how many prime numbers exist between a number, n, and zero**, and describes **the asymptotic distribution of primes**.

$$\lim_{x \to \infty} \frac{\pi(x)}{x / \ln(x)} = 1$$

PRIME NUMBER FORMULA

This formula estimates **the distribution of prime numbers on the number line**.

$$P_n \sim n \ln(n)$$

- P_n = nth prime number
- n = number
- ln = natural logarithm (see section on Logarithms, page 23)

BERTRAND'S POSTULATE

This describes **the gap between consecutive primes**. If you pick any number (n) there will be a prime number p that will be between n and $2n$.

If $n \geq 1$ then there is at least one prime p with the property that

$$n < p \leq 2n$$

PRIME FACTORIZATION AND CRYPTOLOGY

Prime numbers are **used in cybersecurity and encryption**. It's easy to multiply any two large prime numbers to get a number, but difficult to choose any large number and identify what large prime numbers might have been multiplied together to create it.

- A **"public key"** is created, made of the product of two large primes and used to encrypt a message.
- A **"secret key"** consisting of those two primes is used to decrypt the message.
- The public key can be made publicly available, but only you will have the secret key needed to decrypt the message.

CALCULUS

Calculus is the mathematical study and analysis of change. It comprises differential calculus (which evaluates changes over small time intervals) and integral calculus (which evaluates the overall change).

RATES OF CHANGE

Changes such as **population growth** or **temperature change** can be shown on a graph as a **function of time. Finding the gradient of the slope at a specific point along the x-axis helps us understand rates of change.** Rates of change are denoted with the Greek letter "delta": ∂, Δ.

LIMITS

Limits are **used to help us make predictions about what a function might look like at certain points.**

INTERVALS

An interval is a range on a graph between two points on the x-axis. By using infinitely many tiny intervals, **we can more closely approximate how a function changes.**

DIFFERENTIAL NOTATION

$y = f(x)$ or $y = $ a function of x

> The derivative of y (with respect to x) is the change in y with respect to x.
>
> This is denoted as: dy/dx
>
> Put these together and you get:
>
> $$\frac{d}{dx} f(x)$$

INTEGRAL NOTATION

This **allows you to "undo" or reverse a differential.** It shows you the area under a curve.

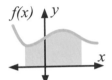
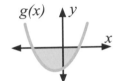
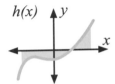

EXAMPLE: POSITION, VELOCITY, AND ACCELERATION

The example of position, velocity, and acceleration can tell us about **the geometric relationships between differentials and integrals**.

- The curve $v(t)$ shows **acceleration and deceleration**, i.e., **changes in velocity.**
- Acceleration is the derivative of velocity: the gradient of the velocity–time curve is the acceleration at a given point in time. $a(1), a(2),$ and $a(3)$ represent acceleration at different times.
- Velocity is the derivative of position: the area between t_1 and t_2 under the curve is the integral which represents **the displacement (change in position)**.

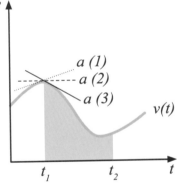

DERIVATIVES AND TRIGONOMETRY

The derivatives of $sin(x)$ and $cos(x)$ relate to each other in the following ways:

The derivative of $sin(x)$ is $cos(x)$

The derivative of $cos(x)$ is $-sin(x)$

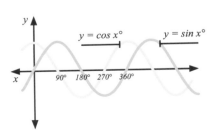

LOGIC

Logic is a methodology for reasoning used to explain ideas, express beliefs, and construct arguments.

- Premises: propositions that **must lead to a conclusion**.
- Propositions: can be **universal** (for all); **particular** (for some); **affirmative** (confirming); or **negative** (refuting).
- Conclusions: **a statement of belief**.

- Validity: when a premise leads to a conclusion it is said to be valid. **Validity is defined by "form" rather than content and is not the same as truth.**

TYPES OF LOGIC

Deductive: a system of syllogistic logic. A **syllogism** is a type of argument that concludes itself through the idea that if the premise is "true" then your conclusions must be "true."

- Premise 1: Daisy is a cow.
- Premise 2: All cows are ungulates.
- Conclusion: Therefore Daisy is an ungulate.

Inductive: operates in terms of **probabilities and *likely* conclusions**.

Abductive: **Depends on available information.** Not all information is available at all times.

Arguments by analogy: Inductive **logic using perceived similarities to infer yet-to-be-observed similarities**.

Reductio ad absurdum: Disproves a statement by **showing it inevitably results in absurdity**.

Paradoxes: Display **valid reasoning from true premises but lead to contradictory conclusions**.

THE BARBER'S PARADOX

Imagine a **barber** who **shaves all and only** those who do not shave themselves. If the barber shaves only people who **do not** shave themselves, does the barber shave himself?

- The barber cannot shave himself because the barber is supposed to shave only people who do not shave themselves.
- But if the barber has not shaved himself, then he should, because the barber must shave those who don't shave themselves.

Let A = people who cut their own hair
Let B = people who do not cut their own hair

In which circle is the barber?

MATHEMATICAL LOGIC

Mathematical logic can be divided into four categories:
1. set theory
2. model theory
3. recursion theory
4. proof theory and constructive mathematics

LOGARITHMS

"Exponential" functions or curves can be expressed as exponential functions that change in accordance to a given formula. They are often described by expressions of "to the power of." Powers are exponents. For example, in x to the power of 2, 2 is the exponent.

A logarithm is the power to which a base must be raised to give a certain number. Logarithmic functions calculate **exponential functions**, which are **inverse functions** (see Calculus, page 21, for more on inverse functions). Logarithms can express large numbers.

- Question: 2 raised to which power equals 16?
- Answer: use the equation "log to base 2 of 16 =" … or log 2 (16) = 4.
- 2 raised to the 4th power equals 16.

THERE ARE TWO TYPES OF LOGARITHM:
- **Log functions** are to base 10 and written as $\log x$.
- **Natural logarithms** are written $\ln x$ and have the irrational number e (≈ 2.718) as their base.

(See Euler's number, page 29, for more on the constant e.)

(Natural log) $\ln N = x \longleftrightarrow N = e^x$

Logarithms simplify multiplication and division:
- $a = b \times c$
- $\log(a) = \log(b) + \log(c)$

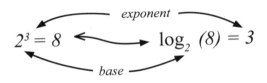

EXPONENTIAL CHANGE

Exponential growth or decay are functions that **become more and more rapid or increasingly slow down over time**.

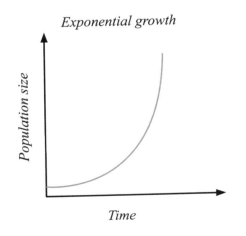

The table below shows the exponential and logarithm laws.

Exponential Laws	Logarithm Laws
$x^a \cdot x^b = x^{a+b}$	$\log(ab) = \log(a) + \log(b)$
$\dfrac{x^a}{x^b} = x^{a-b}$	$\log\left(\dfrac{a}{b}\right) = \log(a) - \log(b)$
$(x^a)^b = x^{ab}$	$\log(a^b) = b \cdot \log(a)$
$x^{-a} = \dfrac{1}{x^a}$	$\log_x\left(\dfrac{1}{x^a}\right) = -a$
$x^0 = 1$	$\log_x 1 = 0$

LOGARITHM HISTORY: JOHN NAPIER AND LOG TABLES

Scottish mathematician **John Napier** spent twenty years calculating **logarithm tables**, which he published in 1614.

PROBABILITY AND STATISTICS

Probabilities measure the likelihood that something will occur. Probabilities work on a scale of 0 to 1: 0 = impossibility and 1 = certainty. Statistics is an area of mathematics focused on the analysis of data. It includes how data is gathered, organized, presented, analyzed, and interpreted.

MEASURING CENTRAL VALUES

- **Mean**: the **average of the numbers in a data set**. To calculate, add up all the numbers and divide by how many numbers there are.
- **Median**: the median is the **"middle" number** of a sorted list of numbers in a data set.
- **Mode**: the **number that appears most often**.

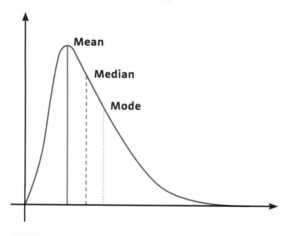

MEASURES OF SPREAD

- **Range**: the **difference between the lowest and highest values**.
- **Quartiles**: to define quartiles, do the following:

1. Put the list of numbers in a data set in order.
2. Divide the list into four equal parts to get quartiles.

- **Interquartile range**: the **difference between the 1st and 3rd quartiles**.
- **Percentiles**: the **value below which a percentage of data falls**.
- **Mean Deviation**: average amount for how far values are from the middle value.
- **Standard Deviation**: **a measure of how spread out numbers in a data set are**. Denoted by the symbol σ (pronounced sigma).

Root mean-square is worked out by **squaring the values, adding them, and dividing by the number of values**.

COMPARING DATA AND CORRELATION

Sets of the same kind of data can be compared by presenting them on the same graph. **The degree to which different data sets match each other is called correlation.** Correlation is measured using numbers 1 to −1 where 1 is perfect positive correlation, 0 is no correlation, and −1 is negative correlation.

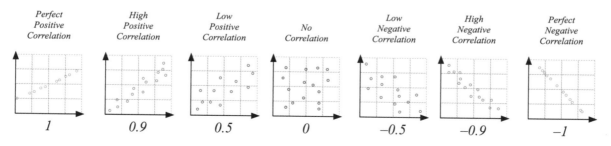

CHAOS

Chaos is the mathematical description of complex systems where the slightest change in initial conditions will dramatically affect the outcome.

Chaos: **An apparently random process taking place within a deterministic system.** A chaotic system is a dynamical (evolving) system whose attractor is fractal.

Attractor: The **equilibrium state** or **numerical point** at which a **dynamic system converges**.

Strange attractor: An attractor is **called strange if it is fractal**.

Fractal geometrics form a structure that is fragmented on all scales of measurement:

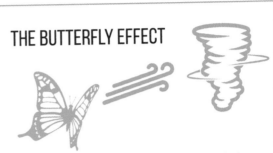

THE BUTTERFLY EFFECT

A principle based on the idea that **a butterfly in the Peruvian rain forest could flap its wings and eventually influence the weather in Glasgow**. This example is often overstated, as other effects as well as butterfly wings might be at work, such as winds over the Atlantic.

*Below: a visualized type of strange attractor called the **Lorentz attractor***

Initial conditions: **Small variations** in the initial conditions of a system can result in **dramatic changes** in the final state of a system.

Deterministic: A system where enough is known about initial conditions and how they change over time to make the outcome **predictable**.

Coupled Pendulums: **Pendulums are deterministic** (see Simple Harmonic Motion, page 34). Coupling two pendulums to create a **double pendulum** (a swing on the end of a swing) results in **erratic motion**. The degree to which they are erratic depends on the height at which the pendulums are released.

Predicting weather
Weather systems have initial conditions that are measurable. To an extent **we can predict the weather, but it's impossible to predict it exactly** as there are too many **changing variables**.

Pseudorandom
An approximated random number often generated by software and hardware. These are **deterministic and never truly random**. When selecting a number truly at random, the number selected **must have an equal and completely unpredictable chance of being chosen**.

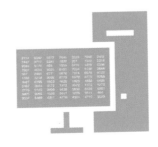

IMAGINARY NUMBERS

Imaginary numbers are also called "complex numbers" and are those that when squared result in a negative number. They emerged from the equation $x^2 = -1$ and are written as a real number multiplied by "i".

"i" is defined by $i \times i = -1$ where i is the square root of a negative number.

$\sqrt{-1}$

Applying **Pythagoras's theorem**, $a^2 + b^2 = c^2$, helps us plot imaginary numbers in the form of **coordinates**. Instead of living along the **number line**, complex numbers exist on the **real plane** or **complex plane**. We can plot coordinates to locate imaginary numbers.

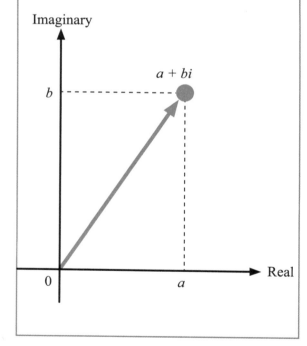

A brilliant and surprising property of i is that, when multiplied out, it results in **four potential values**:

- $i \times i = -1$
- $-1 \times i = -i$
- $-i \times i = 1$
- $1 \times i = i$

The exponents of i can be calculated if the following assumptions are made:

$$i = \sqrt{-1} \quad i^2 = -1 \quad i^3 = -\sqrt{-1} \quad i^4 = 1 \quad i^5 = \sqrt{-1}$$

These results follow a cycle, which is why imaginary numbers have **applications** in **cyclical or oscillating phenomena** and thus vast applications in **signal processing, communication, wireless technologies, imaging technologies, sound analysis, electronics, radar,** and **natural cycles**.

Wherever we find oscillating phenomena, the application of imaginary numbers provides profound insights. Without the use of imaginary numbers there would be no **digital to analog technology,** and no **internet**.

The term "imaginary" was first used in the seventeenth century as a derogatory term for **mathematics that no one understood**.

NON-EUCLIDEAN GEOMETRY

In Euclidean geometry, parallel lines never meet. Different geometries are used in mathematics, and laws regarding parallel lines are not the same in all types of geometry.

Euclidean geometry is carried out on two-dimensional coordinates with no curvature. Non-Euclidean geometry is **the geometry that takes place on any curved surface**, e.g., the surface of a sphere (elliptic) or a surface like a saddle (hyperbolic).

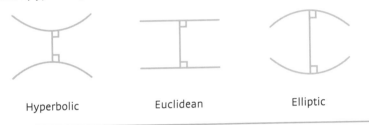

Hyperbolic   Euclidean   Elliptic

ELLIPTICAL PROJECTION

Imagine a plane flying from Beijing to Toronto; from the plane's point of view it is traveling in a straight line, but **it is in fact traveling in a curved path**.

HYPERBOLIC GEOMETRIES

Different models are used in hyperbolic geometry, which is important in **Einstein's theories of relativity**.

EUCLIDEAN PLANE   **SURFACE OF A SPHERE**   **SURFACE OF A SADDLE**

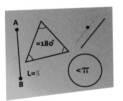  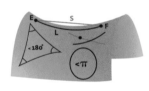

Zero Curvature   Positive Curvature   Negative Curvature
Euclidian geometry   Elliptic geometry   Hyperbolic geometry

Take the internal angles of a triangle for example: **in Euclidean geometry, the internal angles must add up to 180°**. In non-Euclidean geometry (elliptic or hyperbolic), the internal angles do not add up to 180°.

- **Elliptic** = has a positive curvature where **internal angles of a triangle add up to more than 180°**.
- **Hyperbolic** = has negative curvature where **internal angles of a triangle add up to less than 180°**.

Poincaré disk model: projected model of 2-D hyperbolic geometry showing curved lines.

Beltrami Klein model: projected model of curved space onto a 2-D disc resulting in straight lines.

Non-Euclidean geometries help us describe curved surfaces in **electromagnetic and gravitational fields**.

FERMAT'S LAST THEOREM

Pierre de Fermat (1607–65) was a French lawyer who loved mathematics. He was intrigued by an equation similar to Pythagoras's theorem. Instead of limiting the equation to squaring, he wondered if it would hold true when cubed, or to the power of 4, 5, 6… etc.

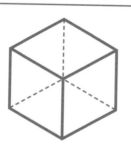

Fermat's theorem is as follows: "**there are no three positive integers x, y, and z for which**:

$$x^n + y^n = z^n$$

for any integer n > 2…"

PYTHAGOREAN TRIPLES

When n = 2 there are an **infinite number of solutions to the problem**. The solutions are called **Pythagorean triples**, where $x^2 + y^2 = z^2$

For example:

$$3^2 + 4^2 = 5^2$$
$$9 + 16 = 25$$

$$161^2 + 240^2 = 289^2$$

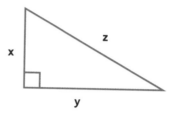

According to Fermat there are **no solutions when x, y, and z are to the power of anything larger than 2**. For example,

$$x^5 + y^5 = z^5$$

has no solutions.

COLLABORATIVE MATH

Fermat thought he could prove this. Annoyingly, he wrote in the margin of a book he was reading that he had a solution to this problem but **"no room to write it in the margin of this book"**… then he died.

Over the centuries, mathematicians have tried to develop a proof for Fermat's theorem.

Math benefits from being **collaborative**; that is, when mathematicians publish their work, other mathematicians check it. While Fermat may have imagined he had found a proof for this problem, many mathematicians believe that he merely *thought* he did, as Fermat worked alone and **no one checked his calculations**.

Many mathematicians have gotten close to proving this theorem, until having it checked and then disproved. In 1994 mathematician **Andrew Wiles** partially proved Fermat's theorem, **paving the way for a full proof** to be developed by others.

Andrew Wiles used **Iwasawa theory** to prove the **Taniyama–Shimura conjecture** in order to demonstrate that Fermat's theorem could partially be proved. **Yutaka Taniyama and Goro Shimura's conjecture (also called modularity theorem)** explores **elliptic curves and connects number theory with topology**. Kenkichi Iwasawa's theory is part of number theory.

EULER'S NUMBER

Euler's number "e" is an irrational number between 2 and 3. It is among the most important constants in mathematics.

$$e = 2.71828182845904523536028747135\overline{27}...$$

Leonhard Euler (1707–83) was a Swiss mathematician who worked extensively on mathematics even after he lost his sight.

The constant e is related to growth appearing in the findings of **applied math** and **physics**; e.g., **population increase** and **temperature change**.

Jacob Bernoulli (1655–1705) was the first to begin working on this when researching **rates of change** and **compound interest**.

Imagine you have one coin in the bank and your bank offers you 100 percent interest at the end of the year; at the end of the year you have two coins. What if that bank offers 50 percent every six months, 25 percent every three months, or 12 percent every month?

Interest earned	Payments per year	Annual total
100%	1	2
50%	2	2.25
25%	4	2.44140625
12%	12	2.61

We could continue this forever, using shorter and shorter time scales, and we would asymptotically approach the value for e. **The constant e is bound up with infinities.**

Another way to calculate e:

$$e = 1 + \frac{1}{1} + \frac{1}{1.2} + \frac{1}{1.2.3} + \frac{1}{1.2.3.4} + \text{etc. ad infinitum.}$$

e AND GROWTH

If we plot $y = e^x$ we get a curve.

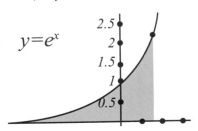

$y = e^x$ is the only function where the **points along the gradient, curvature of the gradient, and area under the curve are the same** (e^x). This relationship is used to make **calculus** (equations that describe rates of change) easier.

THE EULER IDENTITY

π (pi) is an irrational number given by the ratio of the circumference of a circle to its diameter. **Euler figured out an equation that connects π and e.**

$$e^{i\pi} + 1 = 0$$

THE MANDELBROT SET

The Mandelbrot set is famous for its hypnotic beauty. It is a result of complex numbers being used in functions within a defined boundary of the number 2.

To understand how the set works, we start with the **complex plane**,

$$a + bi$$

↑ real numbers ↑↑ $i^2 = -1$

We plot complex numbers like coordinates on the complex plane. The magnitude of the complex number is expressed as $|a + bi|$, which you can see on the graph below:

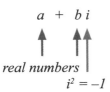

FUNCTION OF z

Imagine a complex number that we will call "c". The function "z" has the following equation. When $c = 1$ and we cycle the results back into $f(z)$, we get the following results:

$f_1(0) = 0^2 + 1 = 1 \rightarrow$
$f_1(1) = 1^2 + 1 = 2 \rightarrow$
$f_1(2) = 2^2 + 1 = 5 \rightarrow$
$f_1(5) = 5^2 + 1 = 26 \rightarrow ...$

The results above are called an "**iteration**" and feed into each other. They show the iterative behavior of 0 in the function z.

The Mandelbrot set is concerned with the size of $|a + bi|$, **of the numbers created by f(z)**, and the **distance they are from the zero coordinate point** on the complex plane. **The iterations in this case are infinitely large.**

If $c = -1$ the results are bound within the number 2:

$f-1(z) = z^2 + -1$
$f-1(0) = 0^2 + -1 = -1 \rightarrow$
$f-1(1) = -1^2 + -1 = 0 \rightarrow$
$f-1(0) = 0^2 + -1 = -1 \rightarrow ...$

SETS OF COMPLEX NUMBERS

If the results of $f(z)$ remain within 2, they exist within the Mandelbrot set boundary and produce fascinating patterns. Within 2 **the function does not diverge when iterated to expand infinitely.**

When we zoom in on a Mandelbrot set, we see **infinite detail and structure**, making it **fractal** in nature.

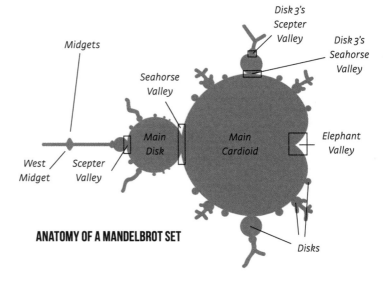

ANATOMY OF A MANDELBROT SET

TOPOLOGY

Topology is the study of different surfaces or spaces. Topology involves the continuous deformation of a shape, without tearing, cutting, or gluing surfaces together.

Topological objects can be **classified depending on how many holes they have**. Topologically speaking, a mug is the same as a doughnut; you can slowly squish a doughnut shape into a mug shape without cutting or tearing it.

TYPES OF TOPOLOGICAL OBJECTS

- **Torus**: This is a donut-shaped object with **one hole**.
- **Double torus**: This is like two doughnuts fused together; topologically speaking it has **two holes**.
- **Triple torus**: This is much like a pretzel and has **three holes**.

MOBIUS STRIP

The **two sides of a mobius strip are continuous with each other;** each side spirals into the other, making it **one continuous surface**. In math-speak, it is described as **a two-dimensional object existing in three-dimensional space**.

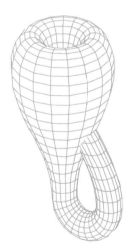

KLEIN BOTTLE

This is a **two-dimensional surface existing in three-dimensional space** where the **external and internal structure of the bottle are continuous**.

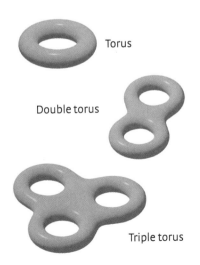

Torus

Double torus

Triple torus

JULES HENRI POINCARÉ

French mathematician **Poincaré** (1854–1912) was a theoretical physicist and philosopher of science. He developed the field of **topology** and contributed to **relativity theory** thanks to his understanding of the mathematics of **spatial distortion**.

The **Poincaré Conjecture** states that **two loops drawn on a torus cannot be continuously tightened to a point**; this makes a torus "not homeomorphic" to a sphere whose loops could be tightened to a point.

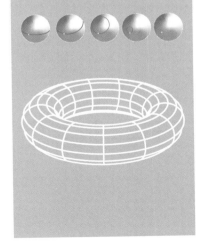

MATHEMATICS

PLANETARY RESONANCE

Celestial mechanics involves combinations of, and interactions between, planetary and lunar orbits, suns, solar systems, and galactic centers. It is about how mutual gravitational influences maintain or alter orbits.

Over time, **orbiting planets and moons exchange momentum and become synchronized**. When the ratios of orbiting bodies relate to each other by whole number proportions, a system is **resonant**.

EXAMPLES FROM OUR SOLAR SYSTEM
- Martian orbit = 687 days, Earth orbit = 365. When working out the ratio between these two numbers, we do not get a whole number; we get 1.88. Thus **Mars** and **Earth** are not in orbital resonance.
- **Pluto** and **Neptune** have a resonance of 2:3.
- The orbits of **Saturn's inner moons** have unstable resonance that don't relate to one another in full number terms; this has given rise to Saturn's **glorious rings**.
- **Jupiter's moons Ganymede**, **Europa**, and **Io** have a resonance of 1:2:4, which means that in the time that Ganymede orbits Jupiter once, Europa has orbited twice and Io four times.

USEFUL MATH: RATIOS AND RESONANT FREQUENCY

To simplify a ratio and work out if two orbits are resonant, multiply or divide both parts of the ratio by a number they are both divisible by (i.e., 3:6 as a ratio can be written as 1:2).

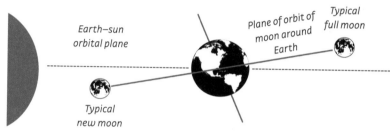

EARTH–MOON

Earth and our Moon are in **"spin-orbit" resonance**, where one side of the Moon is tidally locked to Earth and takes the same amount of time to rotate around the Earth as it does to rotate around its axis. The Moon causes the **tides** on Earth. There are also **solar tides**.

PRECESSION

The Earth has an **equatorial diameter** of approximately 7,926 miles, and 7,900 miles through the poles. It bulges slightly at the equator due to the **orbit of the Moon**, which pulls on the Earth as the Earth pulls on it. The **Moon–Earth gravitational interaction** causes the **Earth's axis** to wobble. This causes the Earth to **"precess"** like a gyroscope.

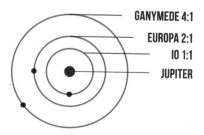

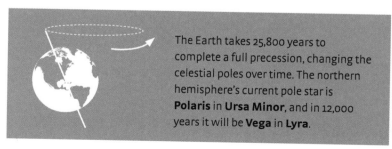

The Earth takes 25,800 years to complete a full precession, changing the celestial poles over time. The northern hemisphere's current pole star is **Polaris** in **Ursa Minor**, and in 12,000 years it will be **Vega** in **Lyra**.

AL-BATTANI

Al-Battani (AD 850–922) was an astronomer and mathematician of the Islamic golden age, most probably from northern Mesopotamia, which is modern-day Turkey.

While little is known about his life, Al-Battani's father is known to have been a maker of scientific instruments.

Al-Battani's work was influential and far reaching. **Copernicus**, **Galileo**, **Kepler**, and **Tycho Brahe** all quoted Al-Battani's writings.

SINES, COSINES, AND TANGENTS

Al-Battani's work stands as some of the earliest evidence of the use of trigonometric functions, which are needed to explain many of the phenomena that scientists seek to understand. He compiled comprehensive tables of calculations for trigonometric functions.

THE KITAB AZ-ZIJ (ASTRONOMICAL TABLES)

Al-Battani was commissioned to create **astronomical tables** that could be **used to predict the movements of the Sun, Moon, and planets relative to the fixed stars**. The tables were used for calculating the **dates of equinoxes**. The original manuscript is kept in the Vatican library. These tables were extremely important for navigational, cultural, and theological purposes.

THE ASTROLABE

An astrolabe is a device developed during the Islamic golden age. Known as an **inclinometer**, it was **used to represent the night sky, with moving parts representing moving celestial bodies**. They could be used to **plot coordinates** and **tell the time**, and were essential for **navigators**.

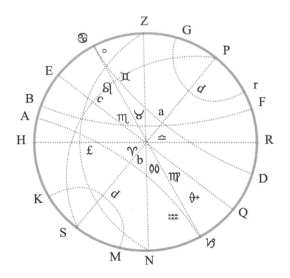

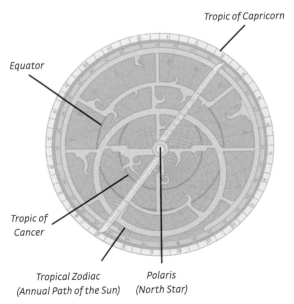

PHYSICS

SIMPLE HARMONIC MOTION

Simple Harmonic Motion (SHM) is used to analyze oscillating motion. It involves the interaction between potential energy (PE), kinetic energy (KE), and angular momentum.

KINETIC ENERGY (K): ENERGY OF MOTION

$$K = \frac{1}{2} mv^2$$

v = velocity m/s
K = kinetic energy in joules

POTENTIAL ENERGY (PE): STORED ENERGY

$$PE = mgh$$

m = mass (in kg)
g = 9.8 N/kg
h = height (in m)

UNIFORM CIRCULAR MOTION

This describes motion in a circle at a constant speed.

The angular velocity equation is:
$$\omega = f \times 2\pi$$

- ω = the angular velocity
- f = force found by Newton's 2nd law
- r = radius

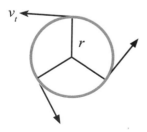

PENDULUM

Pendulum motion can be described using the same math as for **circular motion**. In a pendulum the **angular amplitude** is equivalent to the **radius of a circle**. **At the highest point on the pendulum, the weight stops and changes direction.**

- When it stops, its $KE = 0$ and its PE is at its maximum.
- When the weight is at the equilibrium, KE is at its maximum and $PE = 0$.

SPRING MOTION

- At maximum position on the spring, PE = maximum and $KE = 0$
- Halfway along this motion KE = maximum and $PE = 0$

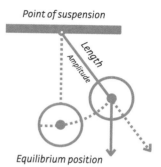

Point of suspension
Length
Amplitude
Equilibrium position

SPRING CONSTANT AND HOOKE'S LAW

Spring-like or elastic materials have a special property called the **Spring Constant. The letter k is used to denote it.**

$$F = -k \times x$$

F = Force in newtons (or the restoring force)
k = The spring constant
x = Amount of extension (measured in meters)

FREQUENCY

The number of full oscillations or cycles per second.

F = wavelength time for one cycle to be completed

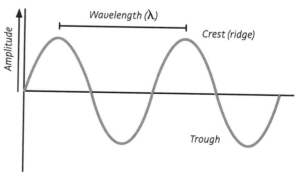

Wavelength (λ)
Amplitude
Crest (ridge)
Trough

PHYSICS

OPTICS

The Ray Model of light is a visual way of explaining how it interacts with different materials.

THE LAW OF REFLECTION

Rays of light incident to (striking) a reflective surface are reflected back at the same angle in the other direction. **Angle of incidence = angle of reflection.**

THE LAW OF REFRACTION

A ray's angle when passing into a substance will be less than its angle of incidence. The angle of refraction is related to the angle of incidence by **Snell's Law**:

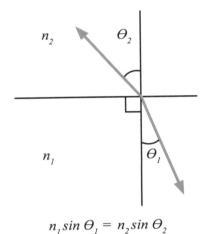

$$n_1 \sin \theta_1 = n_2 \sin \theta_2$$

Angles of refraction are determined by **index of refraction**—an optical property possessed by different substances.

Index of refraction: $n = c/v$

- n = index of refractions
- c = velocity of light in a vacuum
- v = velocity of light in a medium

The higher the index of refraction, the smaller the angle.

LENSES AND FOCAL POINTS

The **power of a lens** can be calculated using this equation:

Power of lens = P
(measured in dioptres)
Focal length of lens = f
(measured in meters)

$$P = \frac{1}{f}$$

CONVEX LENS (CONVERGING LENS) AND CONCAVE LENS (DIVERGING LENS)

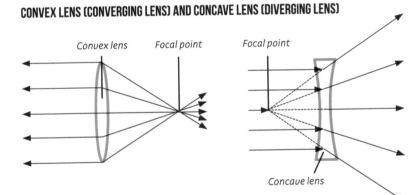

THE LENS EQUATION

$$\frac{1}{f} = \frac{1}{u} + \frac{1}{v}$$

f = focal length (m)
u = object distance (m)
v = image distance (m)

CONSTRUCTIVE AND DESTRUCTIVE INTERFERENCE

Constructive = troughs of two waves collide, or two crests
Destructive = crest + trough cancel each other out

DIFFRACTION PATTERNS

A striped pattern of constructive and destructive interference. Sometimes called **interference patterns**.

SOUND AND ACOUSTICS

Sound is the vibration of molecules in a medium that can be heard (and felt).

TYPES OF WAVE

- **Transverse**: These are **sinusoidal waves**. Light waves are transverse and can travel through a **vacuum**.
- **Longitudinal**: A type of wave that **can travel only through a medium such as the Earth or air**, where energy is transferred by means of **compressions** and **rarefactions**.

This image shows **how a longitudinal sound wave compresses air molecules and distorts atmospheric pressure as it propagates**. The distance between compressions marks a whole wavelength.

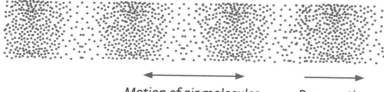

HARMONICS

A fixed string of a given length can be vibrated to whole numbers of **waveforms trapped between two nodes**. The resulting frequencies are **harmonic** with each other; i.e., relate to each other in whole number ratios.

Fundamental
1st Harmonic

First Overtone
2nd Harmonic

Second Overtone
3rd Harmonic

Third Overtone
4th Harmonic

STANDING WAVES: FUNDAMENTAL MODE

Strings on instruments, such as guitar strings, are trapped between fixed points called **nodes**.

Fundamental Mode: Standing Wave on String

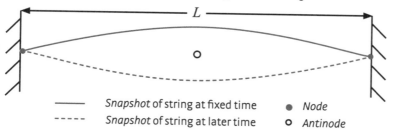

— Snapshot of string at fixed time ● Node
---- Snapshot of string at later time ○ Antinode

RESONANT FREQUENCY

Resonant (or natural) frequency is the frequency that a **string**, **pendulum**, or **elastic object** has when displaced from a stationary position.

DOPPLER SHIFT

When sound is emitted from a moving source, we perceive an **increase in pitch as it moves toward us** and **a decrease when it moves away**. This is a **Doppler shift**, and it happens because the leading edge of oncoming wavelengths are slightly closer to each other when a source approaches, which slightly **compresses the waves and increases the frequency**. The **same phenomena can be observed with light sources**.

Longer wavelength
Lower frequency

Shorter wavelength
Higher frequency

TELESCOPES

For most of human history, vision and imagination have limited our understanding of the universe.

Telescopes were developed from **convex lenses** used for magnification (see Optics, page 35) first described in ***The Book of Optics*** by Islamic scholar **Ibn al-Haytham** (965–1040). His book was translated into Latin and inspired scientist **Roger Bacon** (1214–92) to introduce his methods to **thirteenth-century England**.

TELESCOPES

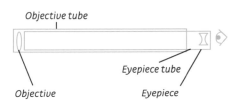

The **earliest known telescope** appeared in 1608 in the **Netherlands.** Astronomers such as Galileo Galilei (1564–1642) used telescopes to look at the night sky and the Moon. His observations proved that we lived in a **heliocentric solar system**.

OUR HELIOCENTRIC OR SUN-CENTERED SOLAR SYSTEM

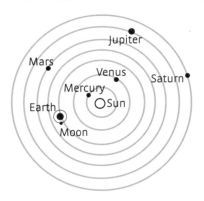

GROUND OBSERVATORIES

- **Uraniborg** was a Danish astronomical observatory and alchemical laboratory used by **Tycho Brahe** (1546–1601).
- The **Jantar Mantar** in **Jaipur, Rajasthan**, completed in 1734, features the **world's largest stone sundial and nineteen architectural astronomical instruments**.

INFRARED OPTICAL TELESCOPES

The **Mauna Kea Observatory** in **Hawaii** houses twelve telescopes built by international astrophysical research groups. One of the largest primary mirrors in the world is that of the **Japanese Subaru** 27-ft. optical-infrared telescope, completed in 1999.

ASTRONOMICAL ARRAYS

- **ALMA (Atacama Large Millimeter/submillimeter Array)**, completed in 2011, comprises sixty-six individual radio telescopes in the **Atacama Desert in Chile** that detect **electromagnetic radiation** at **millimeter and submillimeter wavelengths**.
- The **Spherical Radio Telescope** opened in 2016. Nicknamed **Tianyan**, it is located in **Pingtang, southwest China**.

REFLECTOR TELESCOPES

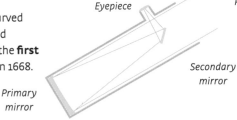

These combine large curved mirrors, flat mirrors, and lenses. **Newton** made the **first reflecting telescope** in 1668.

The **Hooker 100-inch Telescope** (completed in 1917) at **California's Mount Wilson Observatory** helped **Edwin Hubble** prove in 1923 that the **Andromeda nebula** was beyond the **Milky Way**.

PHYSICS

WORK DONE, POWER, AND ENERGY

Whenever a force moves an object, work is done. Lifting, running, walking, climbing, and pushing are all ways in which our bodies "do work."

WORK DONE

Work done includes the **force at work (F)** and the **distance (d) an object is displaced**. Work done is calculated using the following equation and is measured in the same units as energy—joules:

$$\text{work done} = \text{force} \times \text{distance}$$
$$W = F \times d$$

- W is measured in joules (J)
- F is measured in newtons (N)
- d is measured in meters (m)

Energy is the **capacity to do work**, energy is used to do work, **and work done is equivalent to energy transferred**.

ENERGY

Energy cannot be created or destroyed; it is converted from one form into another. **When we convert energy from one form to another, some is ALWAYS lost through dissipation.** Perpetual motion devices are nothing more than an impossible dream.

EFFICIENCY

When energy is transformed from an input to an output, e.g., engine or electrical appliances, the efficiency of the system might need to be known. **Efficiency is calculated as a proportion of the energy supplied to energy output** and is measured in joules (J):

$$\text{efficiency} = \frac{\text{useful energy transferred}}{\text{total energy supplied}}$$

$$\text{percentage efficiency} = \text{efficiency} \times 100$$

$$\text{percentage efficiency} = \frac{\text{useful energy transferred}}{\text{total energy supplied}} \times 100$$

POWER

Power tells us the **rate at which work is being done**. It tells us **how long it takes for energy to be transferred**. It is calculated dividing work done by time, as **it measures work done per unit of time**:

$$P = \frac{W}{t}$$

- P = power, measured in watts (w)
- W = work done, measured in joules (J)
- t = time, measured in seconds (s)

WORK DONE

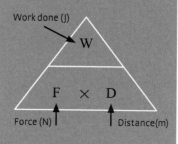

**Work done (J) =
Force (N) x Distance(m)**

The distance must be in the line of the force

KEPLER'S LAWS

German astronomer Johannes Kepler was a mathematician and astrologer. He formulated the Laws of Planetary Motion and wrote a science fiction story called The Somnium. *His wider work led to him being persecuted by Catholic fundamentalists and forced to leave his home.*

MYSTERIUM COSMOGRAPHICUM (THE COSMOGRAPHIC MYSTERY), 1596

Kepler experimented with an idea of the cosmos where three-dimensional polyhedra (shapes) known as the **Platonic solids** corresponded to the six known planets: **Mercury**, **Venus**, **Earth**, **Mars**, **Jupiter**, and **Saturn**. Kepler rejected this idea as it did not match observations.

KEPLER AND TYCHO BRAHE

In 1600 Kepler met **Tycho Brahe**, astrological advisor to **Holy Roman Emperor Rudolph II**. Tycho was an excellent observational astronomer with an excellent observatory. Tycho guarded his measurements closely but was intrigued by Kepler's mathematical expertise. They worked together but argued often. After Tycho's death in 1601, Kepler was appointed as **imperial mathematician**. He had access to the observatory and Tycho's data, which led him to formulate his **three laws of planetary motion**.

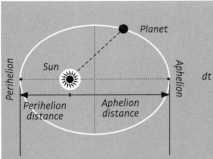
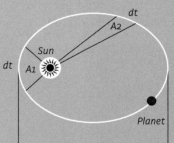
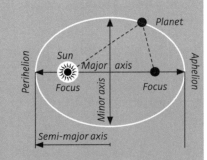

FIRST LAW
All planets move about the Sun in elliptical orbits, having the Sun at one focus.
- An ellipse is a flattened circle or oval.
- The eccentricity of an orbit is a measure of how oval it is.
- Eccentricity is measured between 0 and 1, where 0 means a perfect circle.

SECOND LAW
An orbiting planet around the Sun sweeps out equal areas in equal lengths of time. Area A1 = A2. The second law shows us that in a planet's orbit around a sun it will **speed up when it is closer to its sun**, and travel **slightly more slowly when it is farther away**.

THIRD LAW
This describes **the relationship between a planet's orbital periods and its distance from its sun**.
- The square of the orbital period is directly proportional to the cube of the semi-major axis of its orbit.
- MAJOR AXIS: longest diameter.
- MINOR AXIS: shortest diameter.
- SEMI-MAJOR AXIS: half the longest diameter.

NOETHER'S CONSERVATION LAWS

Conservation laws exist when properties of a physical system stay the same as it evolves over time.

CONSERVATION OF ANGULAR MOMENTUM

- Angular momentum is the product of **rotational inertia** and **rotational velocity** around an **axis**.
- **Total angular momentum** is conserved because it **remains constant unless acted on by an external force**.
- momentum = mass × velocity

Spinning figure skaters increase their rotational speed when they draw in their arms; this decreases angular momentum.

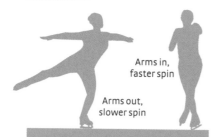

Arms in, faster spin

Arms out, slower spin

Emmy Noether (1882–1935) was a German Jewish mathematician. **Albert Einstein** depended on her extensive mathematical knowledge to help develop his theories of relativity.

Noether's theorem is complicated but in summary states that if there are **symmetries in an equation** then **a physical property is conserved**.

NOETHER SYMMETRY INVARIANCE

One of the most important things about **Einstein's laws** is that **the speed of light stays the same in all frames of reference**, NOT that everything is relative. Many physical phenomena demonstrate conserved properties. The table below shows conservation laws and their **"Noether Symmetry Invariance"**—the property that is preserved.

Conservation law	Noether Symmetry Invariance
Linear momentum	Translational invariance
Angular momentum	Rotation invariance
Mass-energy (E=m)	Time invariance

NEWTON'S CRADLE

Releasing one pendulum at the end of a row of pendulums causes the ball at the opposite end to bounce off, due to the **conservation of linear momentum** or the **symmetry of space**.

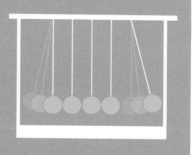

EMMY NOETHER

- Developed **invariance theory** between 1908 and 1911.
- Experienced severe **discrimination** and was fortunate to be wealthy, as she did not receive a salary until 1923.
- Contributed to the **field of topology** from 1920 to 1926.
- **Persecuted** and forced out of her job by the **Nazis** in 1930s Germany.
- Fled for the **United States** to work at **Bryn Mawr College, Pennsylvania**.
- **Died** at fifty-three in 1935 from post-surgery complications.

NEWTON'S EQUATIONS OF MOTION

Newton's laws of motion are used to describe objects in motion. Newton published Philosophiae Naturalis Principia Mathematica *in 1686; it contained his three laws of motion.*

NEWTON'S THREE LAWS OF MOTION

FIRST LAW
Objects will remain in uniform motion in a straight line or at rest, unless acted on by an external force. This is the definition of inertia.

SECOND LAW
The velocity of an object changes when it experiences an external force. Net (overall) force is equal to a change in momentum.

$$F = m \times a$$

- a = acceleration in m/s/s
- m = mass in kg
- F = force in newtons

THIRD LAW
For every action (force) there is an equal and opposite reaction.

NORMAL FORCE

The normal force acts on an object on a surface (such as a table or slide). In accordance with Newton's third law it acts **in the opposite direction of weight and perpendicularly (at right angles) to the surface**.

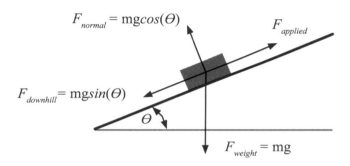

ESSENTIAL CONDITIONS

To understand the science of motion you need to know the following things:
- t = time in seconds
- s = displacement in your position in meters
- v = velocity m/s (rate of change of position measured in meters per second)
- a = acceleration in m/s/s (rate of change in velocity meters per second per second, or meters per second squared, m/s²)

KINEMATIC EQUATIONS

These equations describe the **essential conditions of motion** and how they relate to each other.

$$v = u + at \quad [1]$$
$$s = ut + \tfrac{1}{2}at^2 \quad [2]$$
$$s = \tfrac{1}{2}(u+v)t \quad [3]$$
$$v^2 = u^2 + 2as \quad [4]$$
$$s = vt - \tfrac{1}{2}at^2 \quad [5]$$

SPEED AND VELOCITY

Speed and velocity are defined by the change in distance per unit time.
- Speed is simply the size (**magnitude**) of your rate of change in distance and is a **scalar** quantity.
- Velocity is the same but has **magnitude** and **direction** and is a **vector** quantity.

GRAVITATIONAL CONSTANT

There are many constants of nature in science. The gravitational constant "G" is a constant used to determine the gravitational force between two objects with mass as a result of gravity.

GRAVITATIONAL FORCE

- Gravity **pulls all matter in the universe, even light, toward each other**.
- Gravity **helped to form the universe by pulling gas and dust particles together**.
- The **Sun's gravity holds the planets in their orbits**.
- The **Milky Way galaxy owes its structure to gravity and has a black hole in the center**.

MUTUAL ATTRACTION

If you drop a ball, it falls to the ground because both their masses create a **mutual gravitational force**. The gravitational force between the ball and the Earth is the same, and as the Earth is much bigger, the balls falls toward it.

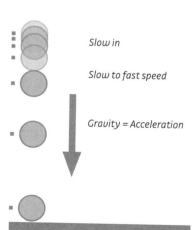

Slow in

Slow to fast speed

Gravity = Acceleration

ACCELERATION DUE TO GRAVITY

The acceleration of a falling object due to gravity is called "G". Using $F = ma$, it is possible to **calculate gravitational force under the influence of gravity** where **force is weight** (which is different from mass).

- On planet Earth the **acceleration due to gravity** G = 9.81 m/s = 9.81 ms^{-1}.
- The **force** is measured in **newtons**, **mass** is measured in kilograms.

The following equation is used to calculate the gravitational force between two masses:

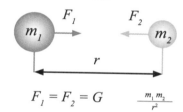

$$F_1 = F_2 = G \frac{m_1 m_2}{r^2}$$

The equation is an **inverse square law** and the farther you are from the gravitational center, the weaker the force.

Inverse square law describes a relationship between physical quantity where a quantity is inversely proportional to the square of the distance from the origin of the force. G is the **gravitational constant**:

$$G = 6.7 \times 10^{-11} \frac{\text{Nm}^2}{\text{kg}^2}$$

Scientist **Henry Cavendish** measured the gravitational constant between 1797 and 1798 using a **torsion balance** made from two lead spheres secured to opposite sides of a horizontal wooden rod suspended using metal wire. Two very heavy spheres were positioned in stationary positions to see if the smaller spheres would be attracted to them. By measuring the degree to which the rod was twisted toward the large spheres, Cavendish calculated the force due to gravity between the masses.

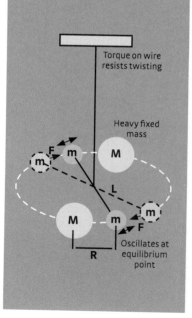

FARADAY AND ELECTROMAGNETISM

Michael Faraday (1791–1867) began his working life as a bookbinder's apprentice. In 1805 he bound a book called Conversations on Chemistry, *written anonymously by a female author named Jane Marcette. This introduced Faraday to science.*

ELECTROMAGNETIC INDUCTION

Faraday discovered **electromagnetic induction**, which was later developed by **James C. Maxwell**.

EMF

Faraday's experiments demonstrated that a changing magnetic field can induce a **voltage**—also called an **electromotive force or EMF**. He observed that **electric fields** have magnetic properties.

ELECTROMOTIVE MOTOR

Due to Faraday's EMF law, the magnetic and electric fields' **mutual interaction** causes parts of the motor to move. A **homopolar motor** is an example of an electromotive motor.

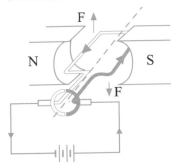

DIY HOMOPOLAR MOTOR

A homopolar motor can be made by carefully bending and balancing a copper wire on top of a battery set on top of a **neodymium magnet**. When positioned properly, Faraday's law **causes the wire to spin around and heat up**.

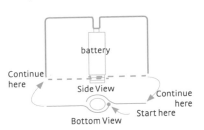

ELECTROLYSIS

Faraday discovered **electrolysis**. When electric current flows through a mixture called an **electrolyte**, the **positive and negative ions** in the mixture can be separated. An electrolyte is a mixture of dissolved ions (charged atoms) and can **conduct electricity**.

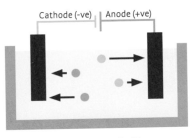

Faraday's law describes that **EMF is equal to the rate of change of the magnetic flux**.
- **Magnetic flux**: The amount of **force** in a **magnetic field**
- **Magnetic field**: The area around a **magnet** where an **electric charge experiences a force**
- **Lenz's law**: States that an **induced current** will oppose changes in **magnetic flux**, resulting in a "pushing" force in the opposite direction of current

FARADAY'S LAW

$$EMF = -N \frac{\Delta \Phi}{\Delta t}$$

(Lenz's Law)

where N = number of turns in copper wire coil
$\Phi = BA$ = magnetic flux
B = external magnetic field
A = area of coil

THERMODYNAMICS

Thermodynamics explores the effects of energy, kinetic theory, and the radiation of thermal energy in large-scale systems. Thermal energy is part of the electromagnetic spectrum and involves the emission of low-energy photons. The four laws of thermodynamics are some of the most important laws in all of physics.

LAW ZERO

If two thermodynamic systems are in thermal equilibrium with a third, then they are in **thermal equilibrium** with each other—they bring each other into equilibrium.

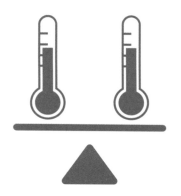

SECOND LAW

The **entropy** (how much the atoms or molecules are able to move around) of an isolated system not in equilibrium will increase over time, moving from **low entropy** to **high entropy**, becoming **increasingly disordered approaching a maximum value at equilibrium**.

Heat flows from a high to a lower temperature; **heat transfer** changes the internal energy of a system.

FIRST LAW

Energy cannot be created or destroyed, it just changes form. The total energy of the universe remains the same. The total energy change equals the net heat supplied to the system minus work done by the system.

$$\Delta U = Q - W$$

Change in internal energy | Heat added to the system | Work done by the system

THIRD LAW

As temperature approaches **absolute zero**, the **entropy of a system approaches a minimum**. In other words, cooling a system to 0 K or −273.15°C or −459.67 °F will cause atoms to stop vibrating.

HEAT DEATH OF THE UNIVERSE

Trillions of years into the future, **all the suns and galaxies and future suns and galaxies across the universe will eventually use up all the atomic hydrogen in the nuclear processes that make stars shine**. Once all the hydrogen in the universe has been used and processed in nuclear reactions, there will be nothing left to burn to make photons. The heat death of the universe describes **the universe's ultimate fate**, where it will evolve to have no thermodynamically free energy to use and will no longer sustain processes that increase entropy.

JOURNEY TO THE HEAT DEATH

Maximum Disorder (thermodynamic equilibruim)

Big Bang

PHYSICS

ABSOLUTE ZERO

Matter and light have a minimal vibrational energy—atoms are constantly jiggling! Absolute zero is the ultimate lowest point on the thermodynamic temperature scale; theoretically at this temperature atoms stop jiggling.

THE KELVIN SCALE

The quantum-mechanical description of absolute zero states that **absolute zero is the ground state of a substance—the point of lowest internal energy.**

LOWEST TEMPERATURE POSSIBLE

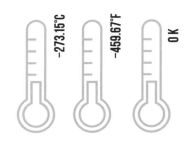

ENTHALPY AND ENTROPY

- **Enthalpy** is the overall amount of energy in a system, which is measured as a change in energy
- **Entropy** is a measure of the disorder of a substance (how much it's able to move around)

DIFFERENCE BETWEEN HEAT AND TEMPERATURE

When a substance or object is heated up, **its molecules have more energy to vibrate with. Heat** is the **total energy of molecular motion. Temperature** is the **average heat or thermal energy of the molecules.**

THE ABSOLUTE ZERO TEMPERATURE SCALE

Absolute zero was first worked out using the **ideal gas law:** $PV = nRT$

$$P = pressure$$
$$V = volume$$
$$n = number\ of\ moles\ (quantity\ of\ gas\ molecules)$$
$$R = the\ ideal\ gas\ constant$$
$$T = temperature$$

SUPERCOOLING AND SUPERCONDUCTIVITY

Scientists have never been able to cool something to exactly 0 K but they've been close, −273.144°C. At this temperature, **substances behave bizarrely**: a whole sample of molecules or atoms will exhibit **quantum effects** such as **superconductivity** and **superfluidity**.

- Some substances such as **hydrogen** and **helium** can be **"supercooled"** and will flow with **zero viscosity**, thus without losing kinetic energy
- **Superconductivity** is a phenomenon of exactly zero electrical resistance

THE MEISSNER EFFECT

The Meissner effect occurs in a superconductor cooled to below its critical temperature T_c. **A mass is able to displace a magnetic field and levitate.**

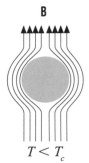

MAXWELL'S EQUATIONS

Scottish physicist James Clerk Maxwell (1831–79) formalized the laws of electromagnetism into unifying equations, bringing together equations describing electric and magnetic fields. He also established, in 1861, that electromagnetic waves were light, further unifying electromagnetism with optics.

ELECTROMAGNETIC SPECTRUM

Maxwell worked out that light was part of this phenomenon and was able to demonstrate that **electricity and magnetism are** in fact **two manifestations of the same phenomenon.**

MAXWELL'S EQUATIONS

$$\nabla \cdot \mathbf{D} = \rho$$

GAUSS'S LAW
This law describes how a **static electric field relates to electric charges**. A static electric field charges points from **positive** to **negative**.

$$\nabla \cdot \mathbf{B} = 0$$

GAUSS'S LAW FOR MAGNETISM
This states that **there are no magnetic charges analogous to electric charges**, and that **magnetic fields emerge due to a magnetic dipole**, which can be represented as loops. This law states that **a magnetic field line that enters a volume must exit that volume.**

$$\nabla \times \mathbf{E} = -\frac{\partial \mathbf{B}}{\partial t}$$

FARADAY'S LAW
This law **describes the phenomenon of electromagnetic induction:**
- The **work** per unit charge needed to move a **charged particle** around a **closed loop** equals the rate of decrease of the **magnetic flux**.

$$\nabla \times \mathbf{H} = -\frac{\partial \mathbf{D}}{\partial t} + \mathbf{J}$$

AMPÈRE'S LAW
Magnetic fields can be generated by the **movement of an electric current**, and magnetic fields can also be generated by **a moving electric field.**

E FIELDS, B FIELDS, AND PHOTONS

One important consequence of his work was showing **how fluctuating electric and magnetic fields travel at the speed of light**. The **magnetic and electric components** of the **electromagnetic spectrum** travel at 90° (or at right angles) to each other.

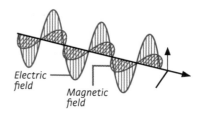
Electric field / Magnetic field

ELECTROMAGNETIC INDUCTION
A **magnetic field** can be produced by **moving an electric current in a coil of wires**, and an **electric current** can be produced by **moving a magnet through a coil of wires**. The magnetic or electric field **vanishes when the current is switched off**.

APPLICATIONS
Maxwell's law can be used to solve specific electric or magnetic potentials in the context of the **boundary value problem**, **quantum mechanical problems**, and in **quantum electro dynamics**. Einstein used this theory to develop his research on **special and general relativity**.

MAXWELL–BOLTZMANN DISTRIBUTION

The Maxwell–Boltzmann equation is used to describe the velocities of molecules in gases at different temperatures. The graph below shows average distributions of speeds of molecules in cold, room-temperature, and hot gases.

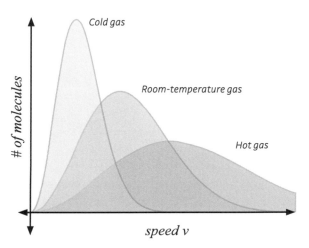

Comparing a hot, a room-temperature, and a cold gas, we can see that **more molecules have a higher speed when a gas is warmed**.

ROOT MEAN SQUARE

In **statistics**, root mean square (RMS) is **worked out by squaring the values, adding them, and dividing by the number of values**. In the **Maxwell–Boltzmann distribution, RMS velocities are used instead of the average speeds**, since the particles are moving in all directions, and if we just used the average velocities some would cancel out.

PROBABILISTIC DISTRIBUTIONS

It is impossible to measure each individual velocity of every molecule; this is why a **distribution of probabilities** best presents **probable speeds, average speeds**, and the **root-mean-square speeds** of particles in gases.

BOLTZMANN'S CONSTANT

$1.38064852 \times 10^{-23}$ m² kg s² K⁻¹

Ludwig Boltzmann (1844–1904) researched **statistical mechanics** and **thermal radiation**.

Boltzmann measured a constant (k_B or k) that represented the **average kinetic energy** of particles in a gas at room temperature and pressure. **Max Planck** was the first person to use it, but he named it after Boltzmann.

Equation for Average Kinetic Energy:

Boltzmann's Constant | k_B | 1.38 x 10⁻²³ JK⁻¹   Gas Constant | R | 8.31 JK⁻¹ mol⁻¹

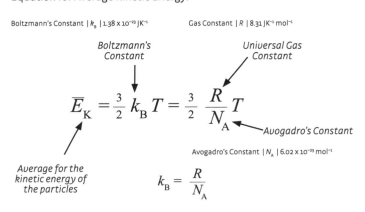

Avogadro's Constant | N_A | 6.02 x 10⁻²³ mol⁻¹

$$\overline{E}_K = \frac{3}{2} k_B T = \frac{3}{2} \frac{R}{N_A} T$$

$$k_B = \frac{R}{N_A}$$

DISCOVERY OF THE ELECTRON

J.J. Thomson (1856–1940) discovered the electron by experimenting with a cathode ray tube in which he measured the charge-to-mass ratio of electrons in a beam.

CATHODE RAY TUBE

A cathode ray tube is a vacuum-sealed tube (almost all the air has been sucked out of it) with a cathode (negative) and anode (positive) on either end that **creates a beam of electrons focused toward the end of the tube**.

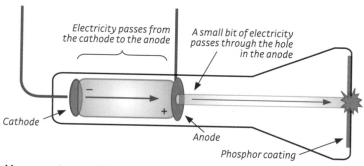

Magnets can be used to **focus the charged beam** and **create images on the phosphor coating**. This is how **particle accelerators** and **old televisions** and **computer terminals** work.

THE CHARGE OF AN ELECTRON

Physicists **Robert Millikan** (1868–1953) and **Harvey Fletcher** (1884–1981) carried out their **"oil drop" experiment** to determine the **amount of electric charge in a single electron**. Published in 1913, it supported the idea that charge comes in **discrete chunks**. They calculated the charge of an electron by suspending tiny charged drops of oil between two metal electrodes. They carefully used charge to balance the droplets against the downward pull of gravity. **The charge of the electron was measured at -1.602×10^{-19} C.** Scientists sometimes use the value of $e = 1$ or $-e = -1$ to denote it.

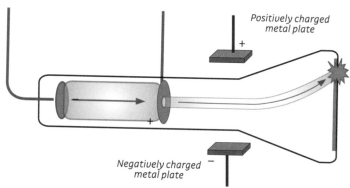

Thomson's discovery led to the discovery of **spectroscopy**.

ELECTROMAGNETISM VS GRAVITY

Gravity is weak compared to electromagnetism. We have the weight of the world pulling us toward the center of the Earth at all times, yet people can still jump. A fridge magnet also has the whole world pulling it down, but electromagnetism can hold it up with ease.

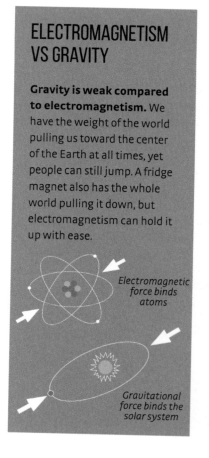

YOUNG'S DOUBLE-SLIT EXPERIMENT

Photons have been shown to have both particle and wave properties. In the seventeenth and eighteenth centuries, the general consensus was that it was a particle; this became the corpuscular theory. Increasing numbers of experiments revealed that photons had wave-like properties.

HUYGENS' PRINCIPLE

Waves exhibit very specific behavior, in that they **oscillate**. Dutch physicist, mathematician, astronomer, and inventor **Christiaan Huygens** (1629–95) figured out that it's **possible to predict where a wave will be in the future by understanding its position when you observe it**.

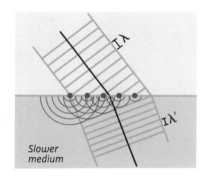

Slower medium

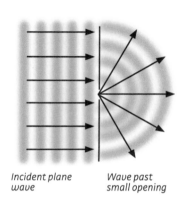

Incident plane wave

Wave past small opening

THE DOUBLE-SLIT EXPERIMENT

The double-slit experiment was first carried out by **Thomas Young** in 1801. He shone a **narrow beam of light** toward a surface with a slit in it, to make the **light continuous**. Behind this was another surface with two thin slits positioned next to each other. A screen was placed on the other side of this, which revealed a resulting **interference pattern**.

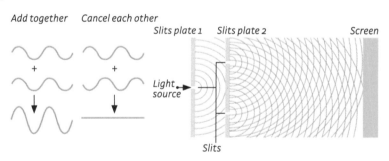

Add together Cancel each other

Slits plate 1 Slits plate 2 Screen

Light source

Slits

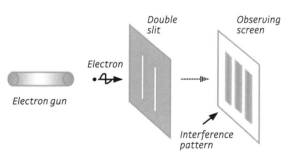

Electron gun

Electron

Double slit

Observing screen

Interference pattern

WAVE PARTICLE DUALITY OF ELECTRONS

The exact same effect can be shown with a **beam of electrons**. If electrons are restricted using a **magnetic field**, and released only one at a time, a **diffraction pattern** is revealed. Scientists **Clinton Davisson** and **Lester Germer** demonstrated this in 1927. Their experiment showed individual **electrons presenting as waves**, the exact position of which cannot be defined as a result. Nor can this result be explained using **classical mechanics**.

THE PHOTON

Light is made of photons, which, like all subatomic particles, are both wave-like and particle-like. Photons oscillate at different frequencies but travel at the same speed—the speed of light.

The speed of light (c) is constant:
$c = 3 \times 10^8$ m/s (in a vacuum)

- The higher the frequency, the smaller the wavelength, the greater the energy.
- The lower the frequency, the larger the wavelength, the lower the energy.

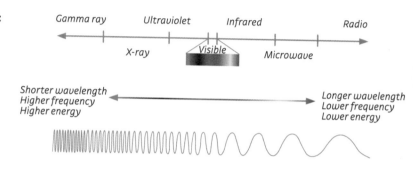

PLANCK'S CONSTANT

Max Planck (1858–1947) was a German physicist who—while heating surfaces that absorb all frequencies of the electromagnetic spectrum (such a surface is called a **black body**)—discovered that light is made of **individual photons**. Planck discovered that **re-radiated light** does not flow evenly from a black body but is **released in discrete packets**.

Planck relation

$$E = h\nu = \frac{hc}{\lambda}$$

where:
E = energy
h = Planck constant
ν = frequency
c = speed of light
λ = wavelength

Planck's constant $h = 6.626 \times 10^{-34}$ Js

PARTICLE PHYSICS

Photons are the force carrier for the electromagnetic force.

MASS-LESS PHOTON

The fact that **photons have no mass** is what makes them **propagate faster than anything else** (in a vacuum).

SPEED OF LIGHT IN WATER

Light interacts with matter. It **reflects** off materials and **refracts** through transparent materials. **It is *not* the case that light slows down** due to being absorbed or bouncing off atoms in a material. Refraction is explained by **quantum mechanics**, where the **incident electromagnetic oscillations** interact with the electrons around the atoms of the material; together they form a **superposition of electromagnetic waves** whose net effect is to slow down slightly.

REFRACTIVE INDEX (z)

Refractive indexes (see Optics, page 35) tell us to which proportion the speed of light through a given material differs from c.

THE RUTHERFORD ATOM

Ernest Rutherford (1871–1937) was a New Zealand–born British physicist. In 1911 he carried out an experiment to understand the inner structure of atoms.

RUTHERFORD'S EXPERIMENT

Rutherford, and colleagues **Hans Geiger** and **Ernest Marsden**, fired a beam of alpha particles at a very thin piece of gold foil in a vacuum.

Alpha particle:

α-particle — Proton, Neutron, 2+

Symbol: $^{4}_{2}He$

Alpha particle is nucleus of helium

A fluorescent (light reactive) screen was used to detect if **alpha particles** traveled through or off the gold foil. The majority of the alpha particles passed straight through the foil, indicating that most of the **atom was empty space**. A few bounced off, and a tiny number were deflected by more than 90 degrees.

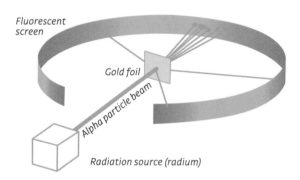

RUTHERFORD'S RESULTS

The experiment demonstrated that **atoms have a small, dense nucleus surrounded by a cloud of negatively charged electrons**.

*This experiment disproved the **Thomson Model**:*

Thomson Model — Rutherford Model

RESULTS

- The **diameter of an average electron cloud** is about 10^{-8} cm.
- The **average diameter of a nucleus** is about 10^{-12} cm.
- The nucleus has an **overall positive charge** and is made of **protons** (positively charged) and **neutrons** (neutral or no charge).
- **Atomic number**: we determine if an atom is a certain element through its number of protons.
- **Atomic mass**: this is the total number of protons and neutrons in an atomic nucleus.

CLOUDS OR ORBITS?

It's not possible to know the position and momentum of electrons at the same time. Sometimes the metaphor of electrons **orbiting** a positively charged nucleus is helpful, and sometimes electrons are best described as a **cloud**. Neither image can tell us exactly what electrons are like.

ANATOMY OF AN ATOM

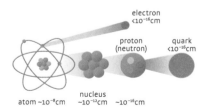

electron <10^{-16} cm
proton (neutron)
quark <10^{-16} cm
atom ~10^{-8} cm
nucleus ~10^{-12} cm
~10^{-16} cm

PHYSICS

MARIE CURIE AND RADIOACTIVITY

Marie Curie (1867–1934) was born in Poland and moved to France to study physics. She received the 1903 Nobel Prize for the codiscovery of radiation with her husband, Pierre. She was the first woman to win a Nobel Prize, and the first person and only woman to win twice.

EMISSION OF RADIOACTIVE PARTICLES

Radioactive atoms will emit **"ionizing energy"** in the form of different particles.

- Alpha particles: helium nucleus—two protons and two neutrons.
- Beta particles: electron.
- Gamma particles: high-energy photon.

ISOTOPES

Isotopes of an element have the **same number of protons** but **different numbers of neutrons**. They have the same **atomic number** but a different **atomic mass** (see The Periodic Table, page 87, and Carbon Dating, page 88, for more).

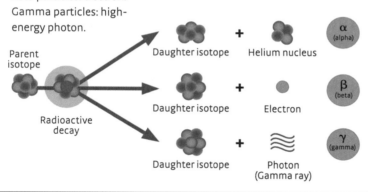

TRANSMUTATION

Radioactive atoms and isotopes are **unstable**, so they will emit particles, **transmuting** into different atoms until they reach **equilibrium**.

$$^{235}_{92}U \xrightarrow{\text{decay by releasing an alpha particle}} ^{4}_{2}\alpha + ^{231}_{90}Th$$

$$^{14}_{6}C \xrightarrow{\text{decay by releasing a beta particle}} ^{0}_{-1}\beta + ^{14}_{7}N$$

$$^{235}_{92}U \xrightarrow{\text{energy shed through gamma wave}} \gamma + ^{235}_{92}U$$

RADIOACTIVE HALF-LIFE

Radioactive half-life (decay) is a random process that happens exponentially. Half-life is the rate of decay of a radioactive isotope. What is meant by decay is the **rate at which radioactive isotopes transmute into a more stable nucleus**.

The particles emitted by radioactive atoms will penetrate or be absorbed by different materials to different extents. Radioactivity is **dangerous** because these particles are **energetic** and **damage DNA** when they **penetrate cells**. Marie Curie was not aware of how dangerous radiation was.

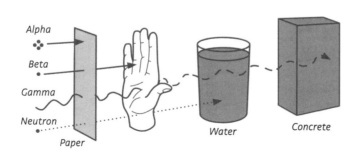

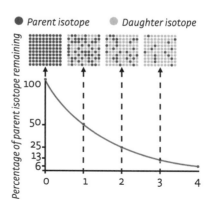

THE PHOTOELECTRIC EFFECT

Light waves can be observed to knock electrons off a metal plate connected to a circuit, causing them to fly across an open space and arrive at another plate, where they complete the circuit.

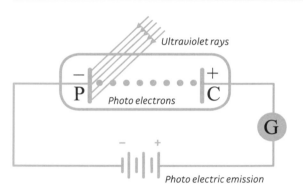

MAKING SENSE OF OBSERVATIONS

1. For electrons to have **enough energy to escape the surface**, the **light striking the surface** must have a **minimum frequency**. No electrons are produced if the frequency of the light waves is below a critical value.
2. The energy of the electrons **doesn't depend on the intensity of the light** striking the surface.
3. Once light reaches the surface, the electrons are emitted immediately.

EINSTEIN'S SOLUTION: THE PARTICLE NATURE OF LIGHT

In 1905 **Einstein** published two papers: one on **relativity**, and one that used **Planck's Constant** (h) to explain the **photoelectric effect**. Einstein's theory was able to predict that the **maximum energy of ejected electrons should increase linearly with the frequency of the applied light**.

$$E = h\upsilon$$

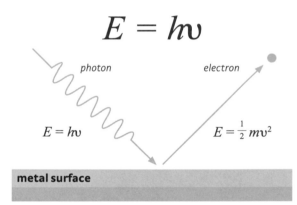

E = MC²

In Einstein's 1905 paper he published **one of the most famous equations of all time**. The paper explained the **mass–energy equivalence equation**.

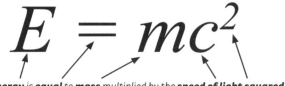

Energy is **equal** to **mass** multiplied by the **speed of light squared**

NEW DAWN

The discovery of the photoelectric effect marked the dawn of quantum mechanics, where specific frequencies of light enabled the release of a photoelectron in a circuit rather than the intensity of light in a fixed frequency.

GENERAL AND SPECIAL RELATIVITY

Einstein's theory of special and general relativity explains the relationship between space and time. The more massive the object, the greater the gravitational distortion of space-time.

FRAMES OF REFERENCE

- Frames of reference in motion **move relative to each other**.
- The **laws of physics are the same in all** frames of reference.
- The **speed of light is the same for all** frames of reference and is **independent of the motion of the source**.

SPECIAL RELATIVITY

Special relativity applies only in frames of reference that are **not accelerating**.

GENERAL RELATIVITY

This theory explains that gravity arises from the **curvature of space and time** due to an object with mass (a massive object). The more massive the object, the bigger the distortion in space-time.

RELATIVISTIC MASS

The greater an object's speed, the greater its momentum. **As it approaches the speed of light, its energy and momentum increase asymptotically.**

RELATIVISTIC MASS INCREASE

$$m(v) = \sqrt{1 - \frac{v^2}{c^2}}\, m_0$$

GRAVITATIONAL LENSING

Scientist **Arthur Eddington** (1882–1944) conducted an experiment during a **solar eclipse**, observing that the **light from distant stars** was being **distorted by the Sun's mass**. Light has no mass and travels in a straight line; **distortions in space-time cause it to bend**. This is gravitational lensing.

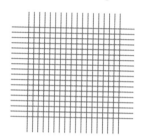

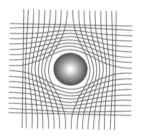

LENGTH CONTRACTION— THE LORENTZ CONTRACTION

A **moving object's length** is measured to be **shorter than its length measured when at rest**. As an object's speed approaches c, its length gets shorter.

$V = 0$ $V = 0.3C$ $V = 0.6C$ $V = 0.9C$

Velocity Increase →

$$L = L_0 \sqrt{1 - \frac{v^2}{c^2}}$$

TIME DILATION

Light traveling within a moving frame of reference with speed "s" does not travel at speed $c + s$. **The speed of light is the same in all frames of reference**, resulting in time dilation (see Time Dilation, page 80, and GPS, page 160, for more).

$$T_0 = \sqrt{1 - \frac{v^2}{c^2}}\, T$$

SCHRÖDINGER AND THE WAVE EQUATION

Erwin Schrödinger (1887–1961) was an Austrian quantum physicist who developed an equation called the wave function that calculates the probability of a particle's position.

WAVE-PARTICLE DUALITY

All subatomic particles exhibit **wave and particle behavior**.

THE COPENHAGEN INTERPRETATION OF QUANTUM MECHANICS

From 1925 to 1927, **Niels Bohr** and **Werner Heisenberg** put forward an interpretation of **quantum mechanics**:
- Particles do not have definite properties until they are measured;
- We can only predict the **probability distribution** of a particle;
- The **act of measurement affects the system**.

SUPERPOSITION

Particles are **simultaneously waves and particles**, with properties spread across space or concentrated at one point depending on how they are measured. They are a **superposition** (combination) of waves.

(Wave function) Ψ

$p = \hbar k$ *(particle momentum = reduced Planck constant x wave vector)*

WAVE FUNCTION

The wave function tells us the probability of finding a particle at a particular position. It predicts the **eigenstate*** of a quantum object, not the exact position.

*Eigenstate: a wave function in a superposition of eigenstates, until we carry out the calculation to find an "eigenvalue" with a precise position.

RADIAL PROBABILITY

Probability of finding electron vs. Distance from nucleus (a_0 = most probable distance)

1s, 2p, 2s, 3d, 3p, 3s at $5a_0$, $10a_0$, $15a_0$, $20a_0$, $25a_0$

Radial probability: the probability of an electron being found within a specified region away from the nucleus

SCHRÖDINGER'S CAT

In 1935 **Erwin Schrödinger** imagined an experiment to test the **Copenhagen interpretation of quantum mechanics**.
- Imagine a flask of poison, a radioactive source, and a detector in a sealed box, arranged such that if the detector detects radioactivity then a mechanism is activated that releases the poison—killing any living thing in the box.
- If a cat is in the box, then, due to the **quantum superposition** of the radioactive particles (which as **quantum objects** can be in a superposition and thus in several places at once), the radioactive source can both trigger and not trigger the chain of events that lead to the poison being released.
- By **observing** (opening the box), the **wave function collapses** and the cat lives or dies.
- A cat cannot be in a superposition of states. Schrödinger's point was to illustrate the **disparity between quantum and classical physics**.

PHYSICS

THE UNCERTAINTY PRINCIPLE

The Heisenberg Uncertainty Principle states that it is impossible to simultaneously measure the exact position and momentum (thus speed) of a particle.

Heisenberg's principle can be explained by **mathematical relationships** describing the **constraints in measuring position and momentum simultaneously**:

$$\Delta x \cdot \Delta p \sim \hbar$$

delta-x times delta-p is proportional to h-bar

The principle can also be used to describe constrains on measurements of time and energy:

$$\Delta E \cdot \Delta t \sim \hbar$$

delta-E times delta-t is proportional to h-bar

- h-bar: "reduced **Planck constant**" (Planck's constant divided by 2π)
- delta-x: **uncertainty** in **position** of a particle
- delta-p: uncertainty of **momentum**
- delta-E: uncertainty of the **energy** of an object
- delta-t: uncertainty in **time measurement**

QUANTUM INTERPRETATIONS

There is a difference between what quantum mechanics does and what we think quantum mechanics means. **Richard Feynman** said, **"everything that can possibly happen in quantum mechanics does happen,"** stating that quantum paths take all possible routes through space. In **classical physics** we do not need to question what is meant by "path" or "force," but in **quantum mechanics** we do.

There are many *interpretations* of quantum mechanics; six are listed below. Some interpretations depend on **wave function collapse**, while others do not. Interpretations where wave function collapse is required include:

- the **Copenhagen interpretation**;
- the **transactional interpretation** (TIQM);
- the **von Neumann interpretation**.

Interpretations where collapse is not considered or is considered an optional approximation include:

- the **Broglie-Bohm interpretation**;
- the **many-worlds interpretation**;
- the **ensemble interpretation**.

WHAT DO WE KNOW?

The **mathematics of quantum mechanics** describes NONE OF THESE interpretations. The mathematics of quantum mechanics states that **quantum systems exist as superpositions**. These interpretations **remain unproven and in some cases unprovable**.

ENRICO FERMI AND BETA DECAY

Enrico Fermi (1901–54) was an Italian physicist who contributed to quantum theory and developed the nuclear reactor. He won the 1938 Nobel Prize in Physics for discovering how to induce radioactive reactions.

In 1934 Fermi developed a **theory of beta decay**, which included **Wolfgang Pauli's** idea of **the neutrino. This was initially presumed to be massless and chargeless**, but we now know the neutrino has a **mass**.

During radioactive decay, **a radioisotope will emit particles and energy to become stable**. It could emit a number of different particles, including an **alpha particle** (two protons and two neutrons), a **beta particle** (electron or positron) **with neutrino**, or only a neutrino, or a **gamma particle**.

BETA DECAY

Beta decay takes place when a **neutron decays**. When a neutron (in an **atomic nucleus**) decays, it turns into a **proton** and an **electron**. Beta decay can also be the emission of a **positron** (**antielectron**) and an **electron neutrino**.

Beta minus decay in atom "A" results in the emission of an electron (beta minus) plus electron neutrino and new isotope "B."

Beta minus decay in atom "A" results in the emission of a positron (beta plus) plus electron neutrino and new isotope "B."

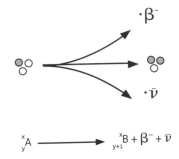

$$^x_y A \longrightarrow \, ^x_{y+1}B + \beta^- + \bar{\nu}$$

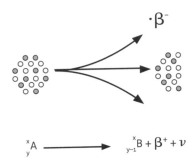

$$^x_y A \longrightarrow \, ^x_{y-1}B + \beta^+ + \nu$$

COSMIC RAYS

Charged particles such as protons travel across the vacuum of space, **emanating from the Sun and stars**, which shine due to **nuclear processes**. When these particles approach Earth, they change direction in the Earth's **magnetic field** and bump into atoms in the Earth's **atmosphere**, resulting in **sprays of many different particles**. Cosmic rays are a **natural phenomenon** and **continuously shower through the atmosphere**. Understanding beta decay has contributed toward our understanding of neutrinos and particle interactions.

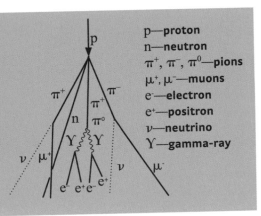

p—proton
n—neutron
π^+, π^-, π^0—pions
μ^+, μ^-—muons
e^-—electron
e^+—positron
ν—neutrino
Υ—gamma-ray

ELECTRON STATES, QUANTUM NUMBERS

Quantum spin is nothing like the spin of a spinning top. Spin is the intrinsic angular momentum of a particle. It gives particles magnetic properties.

- Particles with spin are like tiny magnets with a **north and south pole**.
- Magnetic moment is the **size and direction of the field**.

MAGNETIC MOMENT

Whenever a charged particle moves, it creates a **magnetic field**—we see this in **electromagnetic induction**. Magnetic moment happens due to **spin**; given that **quantum objects** are best described by the **wave function**, this is very strange. The idea of spin arises from the mathematics that describe what quantum objects actually do.

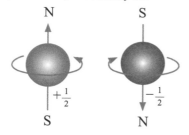

LEFT-HANDED AND RIGHT-HANDED SPIN

Particles can spin **clockwise** in the direction of motion (left-handed) or **counterclockwise** in the direction of motion (right-handed)

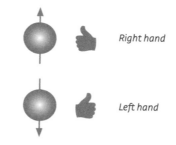

Particles with spin also **orientate themselves in different directions**, so that they **spin up** or **spin down**.

Right-hand particle Spin up

Left-hand particle Spin up

Right-hand particle Spin down

Left-hand particle Spin down

QUANTUM NUMBERS

Quantum numbers describe the **orientation of a particle**. For an **electron**, they describe this orientation in an **orbital**.
- **Fermions** have a **half integer spin** of $+\frac{1}{2}, -\frac{1}{2}$.
- **Bosons** have an **integer spin** of 1, −1, and follow the **Pauli exclusion principle**.

SPIN AND ELECTRON ORBITALS

Electrons around atomic nuclei arrange themselves in **specific 3-D shapes** according to the **wave function**. They change depending on **how close they are to the nucleus and other electrons**. Electrons fit around the nucleus according to the **aufbau principle**, filling energy levels one by one, then pairing up where one has spin $+\frac{1}{2}$ and the other $-\frac{1}{2}$.

ELECTRON CONFIGURATIONS

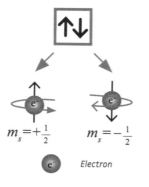

ELECTRON CONFIGURATIONS OF NITROGEN, OXYGEN, FLUORINE, AND NEON

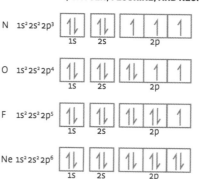

DIRAC AND ANTIMATTER

British scientist Paul Dirac predicted the existence of antimatter. The Dirac equation is a wave equation derived in 1928 that describes spin-½ massive particles (particles with a nonzero mass when at rest).

It was the first equation to hold together in **both quantum mechanical and relativistic contexts**. Dirac won the **Nobel Prize** in 1933 and predicted the existence of **neutrinos**.

$$(i\partial\!\!\!/ - m)\psi = 0$$

The Dirac equation is a "relativistic wave equation." It predicted the existence of antimatter.

THE DIRAC EQUATION

The equation $x^2 = 4$ has **two possible solutions** ($x = 2$ or $x = -2$). This was important in Dirac's discovery. His equation described **electron energies**, with one solution positive and the other negative. Dirac suggested that the negative sign must indicate **antiparticles**.

- The antiparticle of the electron is the **antielectron**, known as the **positron**.
- The positron has **opposite charge** to that of an electron; the electric charge of a positron is +1 e.
- The charge of an electron or positron is a constant of nature denoted by the letter **e**. A positive (+) or negative (−) depending on the type of charge.
- Positrons have **spin** ½ (which is the same as electrons).
- They also have the same **mass** as electrons.
- The photon is its own antiparticle.

MATTER–ANTIMATTER ANNIHILATION

If a positron collides with an electron, a process called **annihilation** takes place whereby **the particles convert into gamma rays**.

Proton P / Antiproton -P → γ, γ

Electron e / Positron antielectron -e → γ, γ

FEYNMAN DIAGRAMS

American theoretical physicist Richard Feynman (1918–88), who said "All matter is interaction," invented a graphical method of writing particle interactions using pictorial symbols.

FEYNMAN DIAGRAM INSTRUCTIONS

In Feynman's words, "Anything that can happen will happen, if permitted by the laws of physics." Each symbol in a **Feynman diagram** represents a particle going from one place to another.
- The vertical axis is the position of a **particle in space**.
- The horizontal axis is **time**.
- Straight lines represent **fermions**: **electrons**, **quarks**, and **neutrinos**.
- Wiggly lines are force-carrying bosons: **photons** and the **W & Z bosons**.
- Loopy lines are often **gluons**.
- Dotted lines are often the **Higgs boson** but can represent the exchange of **virtual particles**.
- The point where lines meet is called a vertex. Each vertex demonstrates events where **conservation laws govern particle interactions**.
- Each vertex must conserve **charge**, **baryon number**, and **lepton number**.
- Arrows denote if something is a **particle** or **antiparticle**—not their direction.

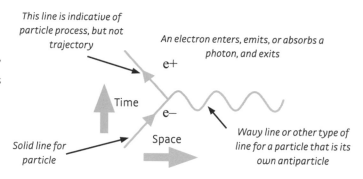

PARTICLE AND SYMBOLS KEY

Incoming fermion
Incoming antifermion
Outgoing fermion
Outgoing antifermion
Incoming photon
Outgoing photon

Fermion
Photon, W, Z
Usually gluons
Usually Higgs boson

BOSON AND FERMION INTERACTIONS

Some particles do interact with each other and others don't. This is why there are **rules** to how Feynman diagrams work. Here are possible particle interactions for the **electromagnetic, weak, and strong interactions**. See The Standard Model, page 62, for more information.

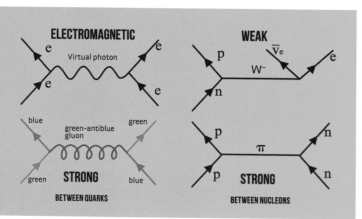

THE MANHATTAN PROJECT

Many scientific discoveries have changed the world for the better, but it's a sad fact that science and technology often emerge from humankind's destructive side.

THE SECOND WORLD WAR AND THE MANHATTAN PROJECT

- **Leo Szilard** and **Enrico Fermi** developed **controlled nuclear reactions** in 1933 and 1934.
- During the **Second World War**, the **Nazis** and the **United States** developed **atomic weapons**.
- **Einstein** supported a letter to **President Roosevelt** from Szilard explaining the atom bomb's power and the dire peril that would ensue if Hitler developed it first.
- The top-secret **Manhattan Project** began in 1939 in the US. It involved more than 130,000 people and the world's leading scientists.
- **J. Robert Oppenheimer** was director of the **Los Alamos lab** where tests took place.
- On July 4, 1945, **Churchill** and the **British government** officially supported using the weapon against **Japan**.
- Einstein sent newly appointed **President Truman** a letter, imploring him not to use the bomb. Truman did not read the letter.
- On August 6 and 9, 1945, Truman detonated two **atomic bombs** on the Japanese cities of **Nagasaki** and **Hiroshima**, instantly killing 250,000 people.

66,000 feet

33,000 feet

Hiroshima 1945 | Nagasaki 1945 | Mount Everest | B83 1970–83 | Castle Bravo 1954 | Tsar Bomba 1961

HARM TO HUMANITY

- The indigenous community called the **Diné** mined **uranium** from the 1940s to the 1980s, experiencing a high incidence of **cancer**. It took the US Government until 1990 to issue the **Radiation Exposure Compensation Act**.
- World powers continued to develop nuclear weapons, **testing them in the world's oceans**, causing permanent **damage to Pacific communities and ecosystems**.
- In 1986 there existed **70,300 active weapons**.
- **The Comprehensive Nuclear Test Ban Treaty** (**CTBT**) was signed in 1996 by 184 countries, ratified by 164.

LEGACY

- In 2018 approximately **3,750 active nuclear warheads** and **14,485 nuclear weapons** existed.
- The **United States** and **Russia** possess more than **90 percent** of the world's nuclear weapons.

THE STANDARD MODEL

The Standard model describes our current understanding of the elementary (or fundamental) particles that make up material reality. It reveals deep descriptions of matter.

FUNDAMENTAL PARTICLES NOT MADE OF ANY OTHER SUBATOMIC PARTICLES

- A particle made up from other **subatomic particles** is called a **hadron**. The proton and neutron are not **fundamental particles** because they are made of **quarks** held together by **gluons**. They are hadrons.
- The **electron** is a fundamental particle as it is not made up of any other smaller particles.
- Electrons are **fermions**.
- Fundamental particles fall into one of two categories: bosons or fermions.

FERMIONS

These have half-integral spin. The **electron** is a fermion. There are two types of fermions: **leptons** and **quarks**. There are three different "generations" of fermions: **electron, muon**, and **tau generations**.

STANDARD MODEL OF ELEMENTARY PARTICLES

Three generations of matter (fermions) / Interactions / force carriers (bosons)

Quarks:
- up (u): ~2.2 MeV/c², charge 2/3, spin 1/2
- charm (c): ~12.8 GeV/c², charge 2/3, spin 1/2
- top (t): ~173.1 GeV/c², charge 2/3, spin 1/2
- down (d): ~4.7 MeV/c², charge −1/3, spin 1/2
- strange (s): ~96 GeV/c², charge −1/3, spin 1/2
- bottom (b): ~4.18 GeV/c², charge −1/3, spin 1/2

Leptons:
- electron (e): ~0.511 MeV/c², charge −1, spin 1/2
- muon (μ): ~105.66 MeV/c², charge −1, spin 1/2
- tau (τ): ~1.7768 GeV/c², charge −1, spin 1/2
- electron neutrino (ν_e): <2.2 eV/c², charge 0, spin 1/2
- mono neutrino (ν_μ): <1.7 MeV/c², charge 0, spin 1/2
- tau neutrino (ν_τ): <15.5 MeV/c², charge 0, spin 1/2

Gauge bosons / Vector bosons:
- gluon (g): 0, 0, 1
- photon (γ): 0, 0, 1
- Z boson: ~91.19 GeV/c², 0, 1
- W boson: ~91.19 GeV/c², ±0, 1

Scalar bosons:
- Higgs (H): ~125.09 GeV/c², 0, 0

BOSONS

- These are responsible for carrying the **fundamental forces of nature**; e.g., photon—which carries the electromagnetic force.
- **Gauge bosons** "exchange" forces in particle interactions and have a spin of 1 (for more on spin see Electron States, Quantum Numbers, page 58).
- **Electromagnetism** arises from the exchange of **photons**.
- The **strong force** is the force that keeps atomic nuclei together and arises from the exchange of **gluons**.
- The **weak force** means atoms can undergo **nuclear fusion**, and arises from the exchange of **W and Z bosons**.

LEPTONS

Leptons do not take part in the **"strong interaction,"** thus do not take part in **gluon exchange**:
- **Charged leptons** have charge $+\frac{2}{3}$ that of an electron.
- **Neutrinos** have zero charge and small masses.

QUARKS

Quarks join together to form **hadrons**, i.e., **proton** and **neutron**.
- **Up-type quarks** (**up, charm**, and **top**) have a charge of $+\frac{2}{3}$.
- **Down-type quarks** (**down, strange**, and **bottom**) have a charge of $-\frac{1}{3}$.

THE WU EXPERIMENT

Chien-Shiung Wu was famous for coming up with the Wu experiment, which demonstrated that the weak interaction of particle physics contradicted the law of conservation of parity.

Chien-Shiung Wu (1912–97) was a Chinese nuclear physicist born in a small town in **Jiangsu province, China**. Wu moved to the United States in 1936. Unfortunately she experienced **racism** and **gender discrimination** in the workplace. She worked on the **Manhattan Project**, researching **beta decay**.

THE WEAK INTERACTION

Chinese physicists **Tsung-Dao Lee** and **Yang Chen-Ning** theorized that in **weak interaction, parity was not conserved**. They worked with **Chien-Shiung Wu** in 1956 on her famous experiment, discovering that the weak force challenges our ideas about **symmetries**, this research winning Lee and Yang the **1957 Nobel Prize**. Many believe that failing to include Wu in the prize was a **great injustice**, as her experiment **fundamentally changed how we understand particle physics**.

PARITY

- Parity (P) symmetry: **parity transformations** concern changes in the direction of the **spatial direction of spin**.
- If P-symmetry is conserved, then no matter which way a particle spins, we should observe the **same results**, and mirror images of particles (with opposite spin) should behave in the same way; i.e., **particles emitted in decays should spin in the same direction as nuclear spin**.

THE WU EXPERIMENT

Wu used **Cobalt 60 atoms** (Co 60) and a very strong magnet to **align their spins, to measure how they decayed**. She observed:

- The **direction of decaying particles** depended on the **direction of spin of Co 60 atoms**.
- The **weak force** acted only on **left-handed matter particles** and **right-handed antimatter particles**.
- The weak force **treated matter and antimatter differently**.
- That **spin (parity) was not preserved in weak interactions**, thus violating P-symmetry.

IF PARITY CONSERVATION HELD TRUE

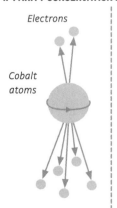

Electrons

Cobalt atoms

EXPERIMENTAL RESULTS

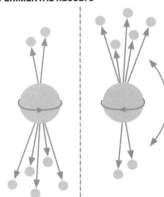

Mirror-image atoms

NEUTRINO OSCILLATION

Neutrinos are leptons, which are fundamental particles. They have a tiny nonzero mass. Exactly how much mass they have is currently unknown. Neutrinos interact only via the weak force.

GHOSTS

Neutrinos do not interact with **mass, electromagnetism**, or **gravity**. They can **travel straight through the Earth and planets**. There are billions of them traveling through us and straight through the Earth every second. Neutrinos can **travel through a mile of lead as easily as a photon does through air**. For this reason, neutrinos are very **hard to detect**.

DETECTING NEUTRINOS

Neutrinos occasionally bump into a **chlorine atom** via the weak interaction. They **exchange a W+ boson with a down quark** and **turn a neutron into a proton**, transforming **chlorine** into **argon**. Neutrinos can also do this with **germanium** and **gallium**.

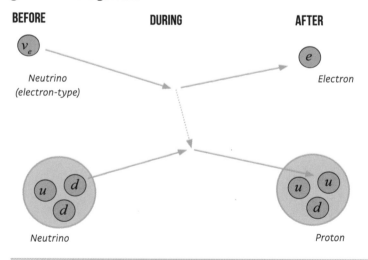

SUPERNOVA (SN) 1987A

In 1987 a **flash of neutrinos** was detected at three neutrino observatories. They had been emitted in a **supernova explosion**, having finally reached Earth after 68,000 years.

NEUTRINO OSCILLATION

There are three different kinds of neutrino: the **electron type, muon type,** and **tau type**. A **neutrino can oscillate into the different neutrino types** as it travels across the cosmos.

CHERENKOV RADIATION

Nothing travels faster than the **speed of light in a vacuum**. However, **neutrinos can travel faster than photons in water**. When a neutrino bumps into a water molecule, a **blue light boom** called **Cherenkov radiation** occurs. Scientists at neutrino observatories JPARK, Japan, and **The SNO, Canada**, use large volumes of **ultra-pure "heavy" deuterium water** (water made with an isotope of hydrogen with one neutron in it).

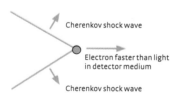

Electrons produced in neutrino detectors can release a burst of blue light called Cherenkov radiation

THE HIGGS BOSON

The Higgs boson gives things mass. Mass is an intrinsic property. The more a particle interacts with the Higgs field the greater its mass.

FORCE FIELDS AND BOSONS

- The four fundamental forces are **electromagnetism**, **gravity**, the **strong force**, and the **weak nuclear force**.
- **Quantum mechanics** shows that the **strength of a force** is the **distribution of particles**: the **more dense the distribution, the stronger the force**.
- **Field** = overall effect of **boson exchange**.
- Proof that the **electromagnetic and weak force are related** was discovered in the 1960s, together forming the **electroweak interaction**; the Higgs boson explains why they are **different but related**.

THE HIGGS BOSON

The Higgs boson is a **vibration in the Higgs field**—just as the **photon is an excitation of the electromagnetic field**.

HIGGS DECAY DETECTION

- The Higgs boson can decay into: the **W and Z bosons**, a **gamma ray photon**, and **quarks (fermions)**. These **decay patterns** demonstrate that the Higgs exists.
- Electron quarks, muon quarks, and tau quarks **interact slightly differently from each other** with the Higgs field.

MASS

Mass is **how much something will resist motion**.

THE HIGGS FIELD

This exists throughout the universe. The **more a particle interacts with it**, the **greater its mass** and the **more it resists motion**. **Photons** do not interact with the Higgs field, have **no mass**, and **travel at the fastest speed possible**.

Peter Higgs (born 1929) and **François Englert** (born 1932) first proposed the Higgs boson in the 1960s, for which they won the **Nobel Prize** in 2013. More than 3,000 scientists from 174 institutes in thirty-eight countries worked on the ATLAS experiment and around 5,000 active people (physicists, engineers, technicians, administrators, students, etc.) worked on the CMS detector at CERN that discovered the Higgs in 2012.

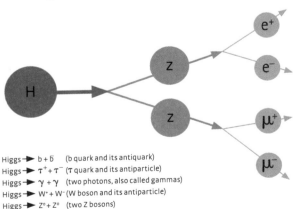

- Higgs → $b + \bar{b}$ (b quark and its antiquark)
- Higgs → $\tau^+ + \tau^-$ (τ quark and its antiparticle)
- Higgs → $\gamma + \gamma$ (two photons, also called gammas)
- Higgs → $W^+ + W^-$ (W boson and its antiparticle)
- Higgs → $Z^0 + Z^0$ (two Z bosons)

QUANTUM ELECTRODYNAMICS (QED)

Quantum electrodynamics is the quantum field theory of the electromagnetic force and describes the behavior of charged particles.

FIELDS

- When forces exert their effect, an **exchange of bosons between particles occurs**. The concept of a field is the **overall effect of boson exchange**.
- **Electrons** are particles; **particles behave like waves**.
- Electrons are an **excitation of the electromagnetic field**, which exists throughout the universe.

VIRTUAL PARTICLES

These are **transient** particles that **blip in and out of existence**. The **uncertainty principle** states that particles can do this on the **Planck scale**. In **QED**, a **virtual photon** is exchanged between **charged particles**.

ELECTRON SCATTERING (REPULSION)

Like charges repel. An electron will repel another approaching electron; this is why the matter we are made of does not collapse in on the empty space inside atoms. On the quantum level, the following is happening:
- Two electrons approach.
- They exchange a virtual photon (one emits, one absorbs).
- This makes them recoil from each other.

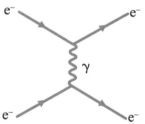

QED

QED is a **relativistic quantum theory** that **describes the discrete quanta exchange of virtual photons in electromagnetic fields** and is relevant to all electromagnetic phenomena, from **magnetism**, to **lightning**, to **electronics**, to **positron–electron annihilation**. It was developed by **Richard Feynman** (1918–88), **Shin'ichirō Tomonaga** (1906–79), and **Julian Schwinger** (1918–94), for which they were awarded the **1965 Nobel Prize in Physics**.

FIELD STRENGTH

The **electromagnetic field and gravity behave similarly**: they have **infinite range** and are **governed by inverse square laws**, meaning their **strength weakens with the square of the distance** (if you double the distance, the force is spread over four times the area).

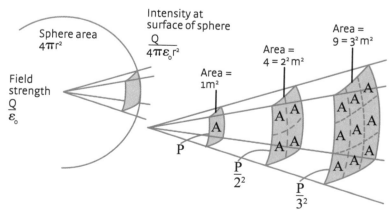

QUANTUM CHROMODYNAMICS (QCD)

Protons and neutrons are nucleons. Nucleons are made of quarks. Quarks stick together because of the strong force, which is mediated by gluons (just as the electromagnetic force is mediated by photons). When quarks interact, they exchange a gluon.

QUANTUM COLOR

Instead of charge (like in an electron), quarks have **"strong force charge,"** also called **"color charge."** **"Quantum color"** has nothing to do with the actual colors we see; rather, color is used to help communicate the **interactions of quarks and gluons**. **Gluons** are also color charged.

INGREDIENTS OF A NUCLEON

- **Proton**: made of two **up quarks** (one is **blue charged** and one is **red charged**) and a **down quark** that is **green charged**.
- **Neutron**: made of two **down quarks** (one **green charged** and one **red charged**) and an **up quark** that is **blue charged**.

Proton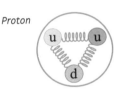

Neutron

THE STRONG FORCE ACTS AT VERY SHORT DISTANCES

The strong force makes **nuclear fusion in stars** possible, **fusing protons** together when their **speeds** bring them **close enough to overcome the repulsion resulting from them both being positively charged**—this is where the **strong force** takes action, at a tiny range of 10^{-15} m. Massive amounts of **energy** are needed to **knock a quark out of a nucleon**; such energy **converts matter into antimatter**, forming a quark—**sprays of matter and antimatter**.

QUARK FLAVORS AND GENERATIONS

- **Flavors**: **up**, **down**; **charm**, **strange**; **top**, **bottom**; and their **antimatter equivalents** (which have different **spin** and **quantum numbers**).
- **Generation**: 1 (**electron type**), 2 (**muon type**), and 3 (**tau type**) tells us the **size of the quark**.

Antiquarks also have color charge = cyan, magenta, and yellow.
- **Antiprotons** are made of **anti-up** (**yellow-charged**), **anti-up** (**cyan-charged**) and **anti-down** (**magenta-charged**) quarks.
- **Antineutrons** are made of **anti-up** (**cyan-charged**), **strange** (**green-charged**), and **down** (**red-charged**) quarks.

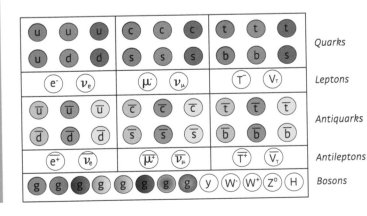

NUCLEAR FISSION REACTOR

Nuclear fission (splitting atoms) occurs in natural radioactive decay and in human-made chain reactions. Radioactive decay exists because of the weak force.

WEAK FORCE

The weak force occurs between **quarks** and **leptons** (**electrons** and **neutrinos**) and involves the **exchange bosons: W and Z bosons**.

- The weak force **alters the properties of particles**, changing them from protons to neutrons.
- W bosons can be **positive** or **negative**, Z bosons are **neutral**; both types are very large.
- It was discovered in 1983 at **CERN**.

THE WEAK FORCE IN ACTION

- A **neutron** is made of **two down quarks** and an **up quark**: $d + d + u$
- A **proton** is made of **two up quarks** and a **down quark**: $u + u + d$

The Feynman diagrams, below, show:
- **BETA DECAY**: a neutron (udd) exchanges a **W⁻ boson** and decays into a **proton (udu), electron**, and **neutrino**; the W⁻ boson carries away the **negative charge in the interaction**.
- **BETA PLUS (POSITRON) DECAY**: proton (udu) decays into a **neutron (udd)** and the exchange of a **W⁺ boson** and **creates a positron and electron neutrino**.

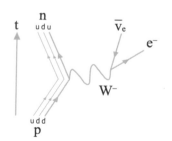 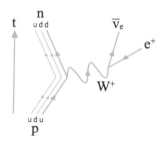

In **nuclear fission, five types of radioactive decay** can happen that **will emit radioactive energetic particles** and **transmute elements** from one kind to another:

- Alpha decay
- Beta negative decay
- Positron emission (also called beta positive decay)
- Gamma decay
- Electron capture

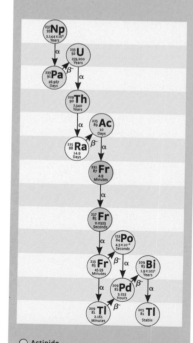

TRANSMUTATION CHAIN FROM ONE RADIOISOTOPE TO ANOTHER

Atomic weapons use **radioactive isotopes uranium and plutonium**. When **uranium-235** absorbs a neutron, it fissions into **two new atoms**, releasing **three new neutrons** and **energy**.

○ Actinide
○ Alkali metal
○ Alkaline earth metal
○ Metalloid
○ Post-transition metal

PARTICLE ACCELERATOR

Particles continuously collide into one another at enormous speeds in space, stars, and the Earth's stratosphere. Scientists can replicate these events using magnets and beams of particles in a high-vacuum environment and study what happens. Particle detectors are sometimes called accelerators.

CLOUD CHAMBER

The cloud chamber was first **invented in 1932 by Carl Anderson** and colleagues and was used to discover the positron (antielectron). A **cloud chamber is a sealed container containing water or alcohol vapor**. If a particle passes through the chamber it knocks electrons off its molecules. This results in a **trail of ionized gas particles** seen as a trail of mist. It is possible to make a simple DIY cloud chamber at home and to observe cosmic rays.

TYPES OF PARTICLE ACCELERATOR

SINGLE-BEAM ACCELERATORS
- Cyclotrons
- Linear accelerators
- Synchrotrons
- Fixed-target accelerators
- High-intensity hadron accelerators (meson and neutron sources)
- Electron and low-intensity hadron accelerators

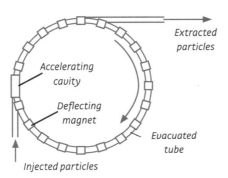

TWO-BEAM ACCELERATORS
- Colliders
- Electron–positron colliders
- Hadron colliders
- Electron–proton colliders
- Light sources

CERN

CERN is the European Organization for Nuclear Research. It is currently the **world's largest particle research facility**. Major discovery: the Higgs boson. Major invention: the internet.

PARTICLE ACCELERATORS AROUND THE WORLD

Particles, i.e., electrons, can be **accelerated by electromagnetic fields**. In a particle accelerator, magnets and a high-vacuum environment are used to do this. There are over 30,000 particle accelerators in operation around the world. Here are just a few:
- Raja Ramanna Center for Advanced Technology, India.
- Synchrotron Light Facility (ALBA), Spain.
- The European Synchrotron Radiation Facility (ESRF), France.
- Centro Atómico Bariloche (LINAC), Argentina.
- The Center for Ion Beam Applications (CIBA), Singapore.
- The High Energy Accelerator Research Organization (KEK), Japan.
- Diamond Light Source and the ISIS neutron and muon facility, UK.
- The Circular Electron Positron Collider (CEPC) is currently under construction.

STARS, THE SUN, AND RADIOACTIVITY

Star formation and evolution are governed by quantum mechanics and gravity.

THE FORMATION OF STARS
- Stars start as **nebulae**.
- **Protons** (H⁺ nuclei) **gravitationally attract but their positive charges repel**.
- This motion increases **kinetic energy**.
- Their **mass attracts** more H⁺.
- **Temperatures** reach 100 million Kelvin.
- H⁺ collide with such force that the **strong force causes fusion**.
- H⁺ fuse into **deuterium**, emitting a **positron**, a **gamma ray**, and a **neutrino**—this is **nuclear fusion**, starting with the **proton–proton chain**.

PROTON–PROTON CYCLE IN STARS

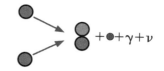

Two protons collided to form a deuterium nucleus, a positron, a gamma ray, and a neutrino.

FORCES INSIDE STARS
- Stars reach **hydrostatic equilibrium**: gravity pulling inward = fusion pushing out.
- Many **fusion reactions** occur; **elements** up to **iron** (Fe) are made; **atoms larger than Fe break apart**.
- When all H⁺ is used up, **hydrostatic equilibrium changes**; the **star expands**, **cools**, and **becomes a red giant**.
- The **surface temperature** of a red giant is about 5,000 K.

MAIN SEQUENCE
- Average stars, like our Sun, are steady-state or main sequence stars.
- After the red giant phase, they shed layers of stellar material, becoming a white dwarf.
- They then cool down, becoming a brown dwarf.

MASSIVE STARS
- When hydrogen burning stops, mass collapses.
- Mass crashes toward the core, causing an explosive supernova—with energy intense enough to produce atoms heavier than Fe.
- This can result in a neutron star.
- In heavier stars, the collapse continues, forming a black hole.

PROTON–PROTON CHAIN

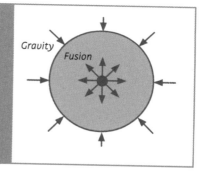

- ν Gamma ray
- γ Neutrino
- ● Proton
- ● Neutron
- ○ Positron

WHAT NEXT?
- Large stars burn hydrogen quickly.
- Smaller stars burn hydrogen slowly.

SOLAR SYSTEMS

Solar systems contain stars orbited by planets and other objects. Some have gas giants orbiting very close and very fast to the star; solar systems with two suns are called binary systems.

The **Earth rotates around the Sun** at an average of 66,000 miles/h. The Sun travels around the center of the galaxy at over 130 miles/s, taking our solar system with it. The planets are **gravitationally bound** to our Sun. Gravitational forces have caused the planets to converge to an **orbital plane** where all the **planets rotate**:

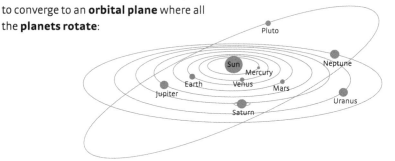

WHAT IS A PLANET?

The **definition of a planet** is **contested**, but some experts suggest that a planet must:
- **orbit** a **star** or **stellar remnant**;
- be **massive enough** to be rounded by its own **gravity**;
- not be undergoing **thermonuclear fusion**;
- and **clear its orbit**.

SOLAR SYSTEM OBJECTS

Objects close to the Sun are smaller and made of dust and rock, while those farther away are dominated by gas and ice. The Kuiper Belt is an outer ring of icy dwarf planets.

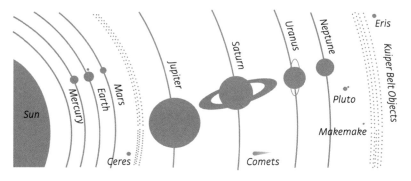

THE FUTURE OF THE SOLAR SYSTEM

As **entropy** increases, less **energy** is available—this is the **second law of thermodynamics** in action. In about five billion years, the Sun will exhaust its **hydrogen**, turn into a **red giant**, engulf the **inner planets**, and **burn out**.

SOLAR SYSTEM FORMATION

All solar systems start as **dust**. Our solar system formed from the **nebula remnants** of a **stellar explosion** (**heavier elements** such as **iron** can only be formed by **nuclear fusion** in stars).

1. Under gravity, nebulae collapse.
2. Dense nebula rotates, flattens, and warms at the center.
3. Debris clumps together to form swirling planetesimals.
4. Largest planetesimals grow; increasing gravity attracts more material.
5. Small planetesimals collide; planets form.
6. Nuclear fission begins at the center of the nebula, releasing energy that blows dust away, leaving a solar system behind.

SPACE OBSERVATORIES

Only a minuscule sliver of the electromagnetic (EM) spectrum is visible to human eyes, but most of what goes on in the universe is invisible to us. Astrophysicists measure and detect photons in the non-visible parts of the EM spectrum such as X-rays, UV, and microwaves.

TARGETS FOR SPACE DETECTION
- Gamma ray
- X-ray
- Ultraviolet
- Visible light
- Infrared and submillimeter
- Microwave
- Radio
- Particle detection
- Gravitational waves

The **Hubble Space Telescope** (HST) was launched in 1990 and is named after US astronomer **Edwin Hubble** (1889–1953). It has four main instruments that detect **ultraviolet, visible,** and **near-infrared radiation.**

Astrosat was **India's first multi-wavelength space observatory,** launched by the **Indian Space Research Organization** (ISRO) in 2015.

X-rays from astronomical sources such as **black holes** and the **center of galaxies** are absorbed by the atmosphere, thus **can be detected only very high in the atmosphere or in space. Supernovas, main sequence stars, binary stars,** and **neutron stars** also emit X-rays.

Chandra X-ray Observatory was launched in 1999 by **NASA**.

Gamma rays (γ-rays) are made in **supernova explosions** and emitted by **neutron stars, pulsars,** and **black holes**. Gamma rays are absorbed by the atmosphere, so high-altitude air balloons or space missions are used to detect them.

The **Dark Matter Particle Explorer** (DAMPE), launched in 2015 by the **Chinese Academy of Sciences** (CAS), detects **high-energy gamma rays, electrons,** and **cosmic ray ions,** in search of **dark matter.**

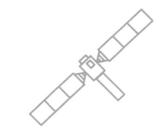

Ultraviolet (UV) rays from light sources **other than the Sun** include **stars** and **galaxies**. Detailed UV observations have helped scientists learn a lot about our own Sun.

Infrared (IR) photons have a lower energy than visible light. Many sources that emit IR rays are **cooler or moving away from us,** including **brown dwarves, stellar nebulae,** and **redshifted galaxies.**

Microwave-detecting telescopes measure the **Cosmic Microwave Background** and energy emitted from the dust of our **own galaxy.**

HUBBLE SPACE TELESCOPE

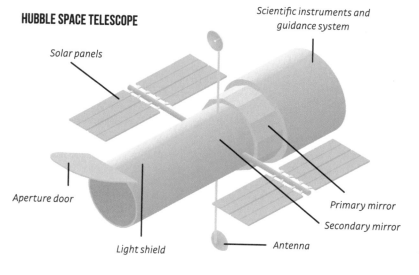

Labels: Solar panels, Aperture door, Light shield, Scientific instruments and guidance system, Primary mirror, Secondary mirror, Antenna

GALAXIES

A galaxy is a system of millions or billions of stars, molecular clouds, and dust, held together by gravitational attraction. Massive galaxies have been observed to contain a supermassive black hole at their center.

GALAXIES

The **orientation** of a galaxy **relative to the arbitrary standpoint of Earth** varies; sometimes only part of a galaxy's structure is visible to us.

CLASSIFICATION SCHEME

HUBBLE SEQUENCE

Edwin Hubble invented a way of **classifying galaxy structure** in 1926 (below).

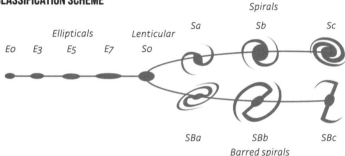

OUR GALAXY— THE MILKY WAY

- 13.51 billion years old.
- **Disk-shaped spiral galaxy** with **two main arms**.
- Contains **hundreds of billions** of **stars**.
- The Sun is **26,000 light years** from the **Galactic Center**.
- **100,000 light years** across, several thousand light years in thickness.
- Has a **bar in the center**, dense with **old red stars**.
- **Supermassive black hole** at its core.

SPIRAL GALAXIES
- Broad flat **spiral disk**.
- Long and thin edge-on, face-on **spiral arms** can be seen.
- Contain **old and young stars**.
- Some have a large **central bulge**.
- Surrounded by a **halo of old stars and gas**.

ELLIPTICAL GALAXIES
- **Ellipsoidal** (squashed ball) shape.
- Consist of small amounts of **dust**, **gas**, and lots of **old stars**.

IRREGULAR GALAXIES
- **Unstructured**.
- Contain **dwarf stars**, **young stars**, and **clouds of** dust.
- Some are too small to have enough **gravity** to form a **structure**.
- Others are the result of **entire galaxies colliding**.

FORMATION

- Galaxies **start off small**.
- Over time, **gravity** attracts more **matter**.
- Gravity overcomes effects of **cosmic expansion**.
- Atoms under gravity form **giant molecular clouds**.
- **Rotation** results in **flat**, **thin disks**.
- **Disruption to rotation** causes **ellipticals**.
- **Major merger**: galaxies of equal mass **collide**.
- **Minor merger**: one galaxy is smaller than the other.
- Generate along **filaments**, **superclusters**, **clusters**, and **groups of galaxies**.

COLLISION/MERGER WITH ANDROMEDA

In about four billion years, our galaxy will collide with the **Andromeda galaxy**.

SPECTROMETRY

The electromagnetic spectrum of light can be used to analyze materials throughout science and medicine. Light is electromagnetic energy: the smaller the wavelength, the greater the energy.

Most of what exists in the universe is **impossible to see with our eyes: molecules, atoms, and light beyond the visible part of the EM spectrum**. The electrons around atoms occupy specific energy levels—distances from the nucleus. When electrons absorb specific wavelengths of energy they **can be excited to different energy levels**. The electrons around each element—i.e., **hydrogen, helium, or iron**—require their own unique frequencies to excite their electrons to different energy levels, and by analyzing the light emitted by distant planets and stars we can determine what they are made of. Elements and molecules **absorb photons of an energy equal to the difference between two energy levels in the atom**.

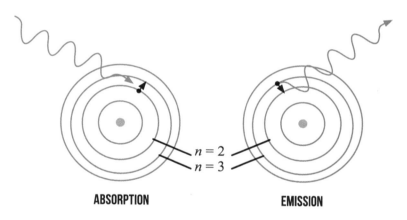

ABSORPTION   **EMISSION**

CALCULATING CHANGE IN ENERGY LEVEL

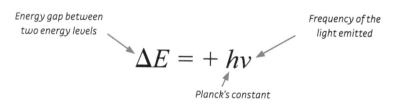

Energy gap between two energy levels → $\Delta E = + h\nu$ ← *Frequency of the light emitted*

Planck's constant

ABSORPTION SPECTRUM

When an atom **absorbs a photon** with an energy equal to the difference between two energy levels, the resulting reflected spectra contains **absorption lines**.

EMISSION SPECTRUM

The energy from the incoming light **can't keep the electron at this higher energy level for long**. When the electron falls back down to the level it was at before absorbing a photon, it will emit it again.

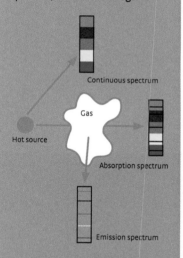

(SOME) APPLICATIONS

Spectrometry can be used for analyzing astronomical distances, composition of stars, composition of the atmosphere of other planets, composition of our own atmosphere, biomedical spectroscopy and tissue analysis, medical imaging, and chemical analysis.

EXOPLANETS

Exoplanets are planets orbiting the stars of other solar systems. Thousands of exoplanets have been discovered orbiting distant suns and millions more are thought to exist.

Some exoplanets have been discovered that are quite **similar to Earth**; some are like the **gas giants** such as **Jupiter**. One exoplant discovered in 2004—called **55 Cancri e**—is extremely **hot**, has a surface made of **graphite** and a thick inner layer of **diamond**!

METHODS FOR FINDING EXOPLANETS

TRANSIT METHOD
The transit method requires measuring **cyclic changes in the light emitted by a star**. If there is a regular dip in **luminosity**, a large exoplanet is orbiting it.

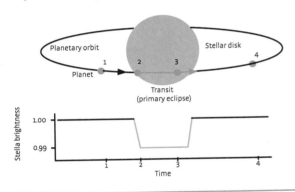

THE DOPPLER SPECTROSCOPY WOBBLE
Most gravitational systems have a **center of gravity** that is not exactly at the center of a star. If a planet's **mass** is significant compared with its star's mass, this offsets the center of gravity, causing a wobble in the star. We can observe this as a slight **Doppler shift** in the light emitted by the star. (See Sound and Acoustics, page 36, to find out more about the Doppler shift.)

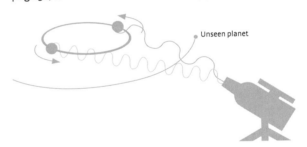

DIRECT IMAGING
This is when an exoplanet is **observed directly** through **telescopes** and **photographic apparatuses**. It's a method that can be applied to finding only relatively **close exoplanets**. It's also limited by the star's glare, which can drown out the light of faint planets.

MICROLENSING
To find smaller exoplanets, **microlensing** is used. This method is based on **Einstein's theory of general relativity**, which states that an object will **curve space-time**, bending the very path along which photons travel. Microlensing measures the distortion due to the bending of space-time to measure masses around distant suns.

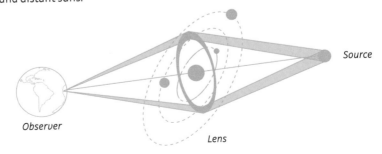

METEORS, ASTEROIDS, AND COMETS

Small solar system objects such as comets, asteroids, and meteors contain materials that tell us about the early universe.

METEORS

Meteors are small, solid pieces of **cosmic debris** from **outer space**, often left behind by comets. When these pieces enter the atmosphere of a planet, such as **Earth**, **friction** builds and the debris becomes **incandescent**, appearing as a brilliant streak of light.

ASTEROIDS

Asteroids are **rocky bodies** of all shapes and sizes that **orbit the Sun**. They have been found between the orbits of **Mars** and **Jupiter**, and in eccentric (very oval) orbits across our solar system. Most known asteroids orbit within the **asteroid belt** between **Mars** and **Jupiter** but they are extremely well spread out. We know of between 1.1 million and 1.9 million asteroids larger than 0.6 mile in diameter, and there are millions of even smaller ones.

THE KUIPER BELT

The **Kuiper Belt** is a circumstellar disk—this means it orbits around a star (our Sun). It is located in the outer solar system reaching from **Neptune** to **50 astronomical units** (**AU**) from the Sun. One AU is 93 million miles. While similar to the **Mars–Jupiter asteroid belt**, the Kuiper Belt is twenty times wider, and contains 20 to 200 times the mass.

COMETS

Comets are celestial objects in **elliptical orbits** and **made of ice and dust**. When their orbits bring them close to the **Sun**, they heat up, releasing a **"tail" of gas and dust particles**, which spray in the opposite direction to the Sun.

The Great Comet

THE GREAT COMET
The **Great Comet** was seen by astronomer **Johannes Kepler** in 1557, when he was just a boy.

HALLEY'S COMET
Halley's Comet is **periodic**, meaning it returns to Earth's about every seventy-five years, and has an **eccentric** (**oval**) **orbit**, as illustrated below. It's next due to visit in 2061.

Halley's Comet

ROSETTA MISSION

The extraordinary **Rosetta mission** of the **European Space Agency (ESA)** landed a spacecraft on **comet 67P/Churyumov–Gerasimenko** in 2014. This mission launched on March 2, 2004, and included a lander module called **Philae**, which was designed to carry out measurements on the comet's surface. Rosetta and Philae used **spectrographs** to gather important data on the nature of the comet.

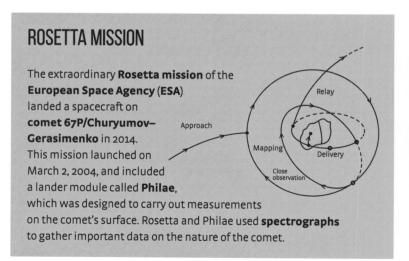

PULSARS AND JOCELYN BELL BURNELL

Jocelyn Bell Burnell is an astrophysicist who codiscovered pulsars.

BELL BURNELL'S RADIO TELESCOPE ARRAY

In 1965 Bell Burnell began her PhD at Cambridge, under her supervisor **Antony Hewish**. She helped to construct a **radio telescope array** in a field near the city to monitor **quasars**.

QUASARS

Quasars (**quasi-stellar objects**) are massive and remote. They emit vast quantities of **radio frequency energy** from both of their poles.

PULSARS

Pulsars are highly magnetized, spinning **neutron stars**. They spin very rapidly, emitting regular pulses of **electromagnetic radiation** in the **radio frequency band**. We can detect the radio pulses only if the pulsars are orientated in the right direction toward our vantage point on Earth.

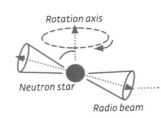

THE DATA FROM PULSAR PSR B1919+21

Bell Burnell noticed **unusual spikes** in her data that occurred every 1.337 seconds. Hewish dismissed them as "human-made" radio sources. Bell Burnell soon ruled this out, deducing that **a new kind of celestial object** had just been discovered.

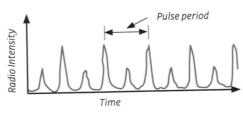

NOBEL PRIZE CONTROVERSY

The **1974 Nobel Prize in Physics** was awarded for the discovery of pulsars; however, Bell Burnell's name was not among those recognized for this achievement. Hewish, her PhD supervisor, was coawarded it **with another male scientist**. Many prominent astronomers criticized Bell's omission, stating that, **even though she was a student, she was the first to observe pulsars** and analyze their unusual data.

JOY DIVISION'S UNKNOWN PLEASURES

Designer **Peter Saville** made use of Bell Burnell's data by stacking continuous layers of data on top of one another for the artwork of British band **Joy Division's** 1979 album **Unknown Pleasures.**

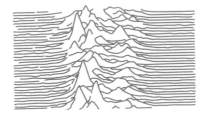

RECENT AWARDS

- 2002–4: President of the Royal Astronomical Society.
- 2008–10: President of the Institute of Physics.
- 2018: Awarded the Breakthrough Prize in Fundamental Physics; **donated the full £2.3 million to help women, ethnic minority, and refugee students become physicists**.

MEASURING THE UNIVERSE

How do astronomers measure the universe? How can we tell it is expanding?

ASTRONOMICAL UNIT

This denotes the **average distance between the Earth and the Sun**. It is used to help astronomers measure large distances. One AU = 93 million miles.

PARALLAX

Parallax results from **viewing a nearby object compared to a distant object from different positions**.

When you are at position B, the tree appears to be in front of the mountain (arrowed).

When you are at position A, the tree appears to be in front of the mountain (arrowed).

Position A   Position B

Parallax is **used to find the distance to stars** by measuring their direction as seen from **opposite sides of Earth's orbit** six months (and 186 million miles) apart. A star 3.26 light years away would show a **"parallax angle"** of 1/3600 of a degree—the parallax of real stars is smaller even than this.

CEPHEID VARIABLES AND STANDARD CANDLES

Cepheid variables are **luminous stars that pulsate regularly**. They are **brighter than the Sun**, and are **used to measure distances**.

American astronomer **Henrietta Swan Leavitt** (1868–1912) cataloged twenty-five **Cepheid variables** in the **Magellanic Cloud** (a **nebula** that orbits the **Milky Way**). Because they were all at roughly the same distance from Earth, she was able to spot a relationship—**the brighter they are, the longer their pulsation period**. This allowed Edwin Hubble to use them as standard candles—objects with a predictable brightness—for measuring cosmic distances.

HUBBLE'S LAW

Edwin Hubble observed a **relationship between the distance galaxies are moving away from us and their speed**: the farther they are, the faster the speed. He calculated the **rate of cosmic expansion** in terms of a constant now known as the **Hubble constant**.

EXPANDING UNIVERSE

We observe the universe as **expanding from our vantage point on Earth**, but the **same is true** of **every vantage point** in the universe as the **space itself is expanding**.

DOPPLER SHIFT

Photons from distant stars are Doppler shifted. When measuring the **Doppler shift**, the following is observed:
- Light from stellar objects is shifted toward **lower frequencies**; **"red Doppler shift"** indicates that distant stars and galaxies are **moving away from us and each other**.
- The more distant the galaxy, the greater the **shift in frequency**.

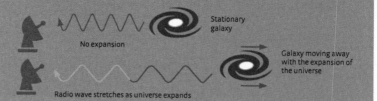

Stationary galaxy — No expansion

Galaxy moving away with the expansion of the universe — Radio wave stretches as universe expands

BLACK HOLES

A black hole behaves like any other object with mass—until you get to the event horizon. A supermassive black hole exists at the center of our galaxy. It is possible to orbit a black hole.

STELLAR EVOLUTION

- If a star's **core** is less than 1.4 times the mass of the Sun, it becomes a **white dwarf**.
- Between 1.4 and 2.8, it collapses into a **neutron star** only thirteen miles across.
- More than 2.8 times, it collapses to form a **black hole**.

BLACK HOLE FORMATION

- **Nuclear fusion** runs out.
- The star contracts due to **gravity**—still emitting **light** at this stage.
- As the **mass collapses, gravity increases** and the **distortion of space-time** becomes so extreme that **light can no longer escape**.
- The **event horizon** is a border beyond which light cannot escape.
- Within the event horizon the black hole, **all mass** and **light collapses into a single point of zero volume**.
- **Particle interactions** take place at the event horizon of a black hole.

SCHWARZSCHILD RADIUS

Gravitational collapse occurs when the **mass of an object** is **compressed into** a volume called its **Schwarzschild radius**.

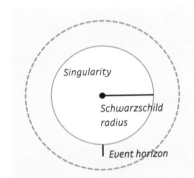

ESCAPE VELOCITY

The **velocity needed to escape the gravitational attraction of a massive body**. The **escape velocity of a black hole is the speed of light**.

OBSERVING BLACK HOLES

Cygnus X-1 is a bright star twenty-three times as massive as our Sun that we can observe orbiting the event horizon of a black hole.

SPACE-TIME DISTORTION

All objects with mass **distort space-time**. The immense mass of a black hole will cause space-time to distort dramatically. **Time slows down at the event horizon.**

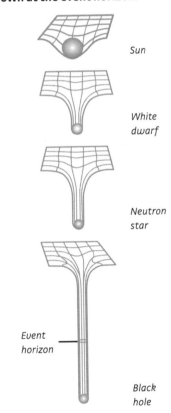

STEPHEN HAWKING

British cosmologist **Stephen Hawking** (1942–2018) combined **quantum mechanics**, **general relativity**, and **thermodynamics** to contribute toward descriptions of black holes.

TIME DILATION

Time dilation occurs as a result of the speed of light being constant in all frames of reference. It is a measurable phenomenon, where the closer an object travels to c, the slower time gets.

- Imagine a spaceship on the Moon traveling at 100 mph ejecting a rocket at 200 mph; an observer sitting on the Moon will see the rocket travel at 300 mph—both speeds added together.
- However, if the spaceship then shoots a laser at the speed of light, "c," although the observer might expect the beam to travel at $c + 100$ mph, this does not happen. **Light always travels at the speed of light**, no matter what its frame of reference.

Imagine two parallel mirrors, A and B, in a frame of reference:

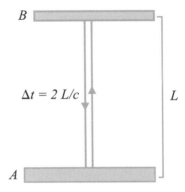

$\Delta t = 2L/c$

An observer at rest views the light beam leave A, bounce off a mirror at B, then return to mirror A having traveled a length of $2L$. They measure the change in time, Δa, as $2L/c$.

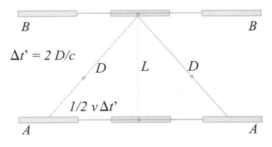

$\Delta t' = 2D/c$

$$t' = \frac{t}{\sqrt{1-\frac{v^2}{c^2}}}$$

t' = change in time
t = rest time
v = velocity
c = speed of light

Einstein's time dilation equation

When viewed by an observer as moving from right to left, light released from A at time $t' = 0$, reaches B at time $t' = D/c$, and is reflected back to A at time $t' = 2D/c$. (t' indicates a different time than our original experiment.) Light travels further in the moving frame of reference and **time dilation** occurs because the **speed of light in a vacuum is always the same**. The **closer your speed is to the speed of light**, the **slower time gets**. The constants control what happens around them.

Time dilation

$$T_0 = \sqrt{1-\frac{v^2}{c^2}}\, T$$

Length contraction

$$L = \sqrt{1-\frac{v^2}{c^2}}\, L_0$$

Relativistic mass increase

$$m(v) = \sqrt{1-\frac{v^2}{c^2}}\, m_0$$

THE TWINS PARADOX

One twin makes a journey on a high-speed rocket while the other stays on Earth. The twin on the journey travels at near the speed of light and returns to the Earth. **Space and time as described by relativity** state that the twin in the rocket is now **younger** than the twin on Earth.

COSMIC MICROWAVE BACKGROUND RADIATION

Any object above absolute zero will radiate electromagnetic waves as thermal energy. The universe is slightly above absolute zero (by about 2.7 Kelvin), meaning that it radiates thermal energy. This is the cosmic microwave background (CMB).

- In 1927 Belgian astronomer **Georges Lemaître** proposed that the **early universe** was hot and dense, and **cooled as it evolved**.
- In 1964 **microwaves were measured to be permeating space**. The **CMB** is the **residual radiation from the Big Bang**.

COBE

The **COBE spacecraft** launched in 1989 measured the **cosmic microwave background**. In 1992 it was announced that there were **tiny variations in the temperature** of the cosmic microwave background. These subtle changes or **"lumpiness"** in the CMB indicated **"quantum fluctuations"—virtual particles in the vacuum of space**.

WMAP

The **Wilkinson Microwave Anisotropy Probe** (WMAP) was launched by **NASA** in 2001 to measure **variations in the CMB and composition of the universe**, which was discovered to comprise:

- 5 percent atoms;
- 27 percent matter that has gravity but emits no light (**dark matter**);
- 68 percent of something that is **pushing the universe apart** (**dark energy**).

ANCIENT LIGHT

The light observed from distant stars and galaxies is **very old** and has **traveled vast distances**. We **cannot observe what is happening "NOW" in distant regions of space**, but only as far back in time as it takes for light to reach us. Light from the **Andromeda galaxy** is 2.5 million years old. CMB radiation was first emitted 13.7 billion years ago—**long before stars or galaxies could even form**.

CMB ON TV

The CMB is the residual heat of the Big Bang. The actual frequency of the CMB can be detected on old-fashioned television sets and can be seen on the "in between" channels on TV sets (called "static"). Atmospheric sources also contribute to static.

COSMIC FOAM

In 2012 a galaxy called MCS0647-JD detected by the Hubble and Spitzer telescopes was observed by astronomers thanks to gravitational lensing caused by a galaxy supercluster—a massive structure made of billions of galaxies.

- MCS0647-JD is the **most distant galaxy currently observed:** 13.3 billion light years away.
- It **seems very young and small** because its **light is so ancient**.
- The **stars in this galaxy** have **already burned up their fuel**.
- Gravitational lensing due to galaxy super-structure shows **filament-like structures** formed by **groups of galaxies**.

COSMIC WEB

This denotes the **large-scale structure of the universe**. **Dark matter** is thought to be creating this structure. **Dust clouds**, **particles**, **supernova remnants**, and **stars** form **filaments of matter** that **connect galaxies** in the universe. This structure is called the cosmic web and contains billions and billions of galaxies.

COSMIC HORIZON
- The **edge of the observable universe** as seen from Earth.
- Distant stars and galaxies at this edge are **unreachable**.
- The **light beyond this point** will **never reach us**.

COSMIC FOAM

Quantum mechanics tells us that **virtual particles** can exist briefly as **fluctuations in space-time**, through what's called the **Casimir effect**. The **vacuum of space** is **rippling with virtual particles**.

DETECTION OF GRAVITATIONAL WAVES

Originally predicted to exist by **Einstein** in 1916, gravitational waves are **wobbles in "space-time"** caused by **large moving masses**. The **Laser Interferometer Gravitational-Wave Observatory (LIGO)** detected **cosmic gravitational waves** by measuring **ripples caused by black holes** merging into each other, winning the **2017 Nobel Prize** for the LIGO team.

LIGO detects distortions in space-time by measuring tiny changes in the position of mirrors that reflect photons in its interferometer.

GRAVITATIONAL WAVES

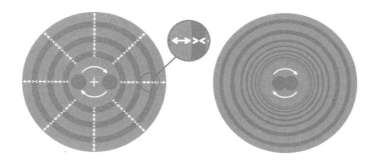

THE BIG BANG

The Big Bang happened 13.8 billion years ago.

EXPANDING UNIVERSE

There is no center to the surface of a sphere. Similarly, there is **no center of the universe; all points are expanding away from one another**. This is difficult for us to understand because we visualize and measure the universe from the fixed vantage point of Earth.

COSMOLOGICAL ERAS

ERA	TIME	TEMPERATURE in Kelvins	ENVIRONMENT
Planck era	10^{-43} seconds after the Big Bang		Forces of electromagnetism, gravity, weak and strong force, are combined.
GUT (Grand Unified Theory) era	10^{-43}–10^{-38} seconds after the Big Bang		Gravity becomes distinct from other forces. Huge amounts of energy released. Space itself inflates from the size of an atom to that of a solar system.
Electroweak era	10^{-10} seconds after the Big Bang		Strong force between gluons and quarks becomes distinct. Electromagnetism and nuclear force are combined. Subatomic particles form, photons present.
Particle era	0.001 second (one millisecond)	Universe expands and cools to 10^{12} K	Particles form, four forces are distinct from one another. Matter and antimatter form from photons and annihilation back into photons.
Era of nucleosynthesis	0.001 seconds–3 minutes	10^9 K	Nuclear fusion possibly forms atoms. Heavier elements begin to form: 75 percent hydrogen, 25 percent helium. Neutrinos, protons, neutrons, electrons. Antimatter is rare.
Era of nuclei	3 seconds–500,000 years	Eventually cools to 3,000 K	Universe is a plasma of particles with free electrons. Photons decouple from matter; light permeates across space. This light is the CMB.
Era of atoms	500,000–1 billion years	3,000 K cools to 2.73 K	First stars are formed, electrons join nuclei to form atoms.
Era of galaxies	To the present day	2.73 K	More structure is formed, galaxies begin to form and evolve.

CHARGE PARITY VIOLATION

CP violation is a violation of CP-symmetry where C-symmetry is charge symmetry and P-symmetry is parity (spin) symmetry. They are said to be violated due to the fact that there is more matter than antimatter in the universe.

PARTICLES AND ANTIPARTICLES

Every particle has its own antiparticle. Some particles—e.g., **photon** and **Higgs boson**—are their own antiparticle. Everything is **made of matter**; cosmologists are curious about why there is **more than one kind of matter than the other. Antimatter** is in every way the **inverse of matter**, sometimes called its mirror image. Matter and antimatter particles have the **same mass but opposite charge and spin**; the **laws of physics** state that these properties should be conserved:

- **Charge symmetry: Interactions** observed **between matter and antimatter** look the **same even if they are oppositely charged.**
- **Parity symmetry: Interactions** observed **between matter and antimatter are unaffected by the "handedness" of a particle**—the direction of spin.
- **Time symmetry: Interactions** look the **same regardless of the direction of time.**

CP VIOLATION

A subtle, **unknown asymmetry between matter and antimatter** is thought to be **behind the fact that there is more matter than antimatter in the universe.**

CPT VS CP-SYMMETRY

When a **pair of quarks are held together by the strong force**, they can **oscillate between red color-charged or blue color-charged** via the **weak interaction** (beta decay). It **takes more time for them to oscillate from blue to red** than **from red to blue**—this is what is meant by **breaking time symmetry**. Similarly, **time goes forward, not backward.**

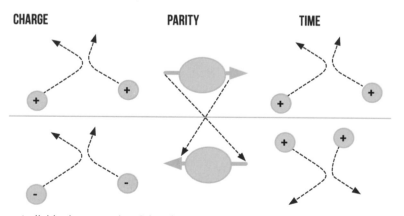

Symmetry
Fundamental symmetry transformation

CHARGE　　　　PARITY　　　　TIME

Individual symmetries violated
Product of all three—"CPT" symmetries—preserved

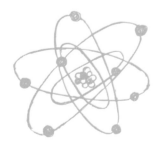

DARK ENERGY AND DARK MATTER

Measurements of the universe reveal that a mere 5 percent of the universe is made of atoms and matter that we can detect—known as baryonic matter. The rest is made of a mysterious kind of matter and energy that we do not understand. It does not react with photons or matter, but is affected by gravity.

- 26.8 percent matter that has gravity **emits no light** (dark matter).
- 68 percent made of something that is **pushing the universe apart** (dark energy).

GRAVITY

Calculations from observing galaxies and galaxy clusters show us that their **mass and gravity** is **not large enough to cause galaxies to form structure. Something is providing extra gravity.**

DARK MATTER

Dark matter **accounts for the extra gravity** needed to give galaxies structure. We **cannot observe it directly** but **we can observe its effects due to gravitational lensing** (mass bends space).

DARK ENERGY

The name given to an **unknown force** that is **causing the universe to expand.** Sometimes discussed as **"antigravity."**

EXPANDING UNIVERSE

- Galaxies in the universe are **accelerating away from one another**.
- The **farther the galaxy**, the **greater the redshift**.
- **More-distant galaxies** are **accelerating faster**.
- Astronomers think that **dark matter and energy** are **behind the Big Bang** and structure and expansion of the universe.

ENERGY DISTRIBUTION OF THE UNIVERSE

- Ordinary (baryonic) matter: 5%
- Dark matter: 27%
- Dark energy: 68%

DARK MATTER CANDIDATES

- Dark matter is **not antimatter or black holes**.
- It is **evenly distributed** throughout the universe and there is a lot of it.
- Particles called **weakly interacting massive particles (WIMPS)** have been hypothesized as a possible dark matter candidate. They react weakly with each other and **baryonic matter**. If they **exist**, they should have a **large mass**.
- Theoretically, they turn up in other **theories that unite quantum physics and gravity**.

MYSTERIES (THE MULTIVERSE, SUPER SYMMETRY, AND STRING THEORIES)

There is a lot we do not know about the universe.

SUPER SYMMETRY

Super symmetry is a property of a theory such that **force and matter are treated equally in an equation**. The **Standard Model** explains a lot, but it is **currently incomplete**. Super symmetry could be a property of the Standard Model.

GUT: abbreviation for Grand Unified Theory

TOE: abbreviation for Theory of Everything

UNITING FORCES

- The behavior of the forces of nature converge at a **very high energy**.
- The discovery of the **Higgs** boson showed us that the **electromagnetic and weak forces are both aspects of an electroweak force**.
- A **GUT** must describe **how the electroweak and strong force can be unified**.
- Perhaps all four forces of nature could be **different manifestations of an all-inclusive single force**.

QUANTUM GRAVITY

Currently, gravity is a **geometric theory of space**. Gravity and how it works on the **quantum scale** has **yet to be included in the Standard Model**. The **graviton** has been **hypothesized to be the force carrier for gravity**—there is, **as yet, no evidence of its existence**.

GUT AND TOE ENERGY EXTRAPOLATION

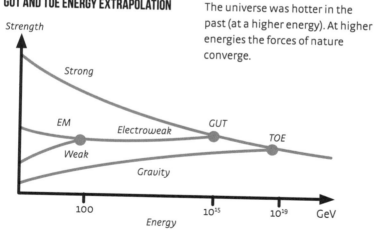

The universe was hotter in the past (at a higher energy). At higher energies the forces of nature converge.

MODIFIED THEORIES OF GRAVITY

Einstein's theories explain cosmological phenomena very accurately, but the **expansion of the universe and dark energy are a mystery**. Some astronomers suggest that **subtle modifications in general relativity need to be made**.

MULTIVERSE

The **"many worlds" interpretation of quantum mechanics** suggests that **multiple realities are formed when the wave function collapses**.

STRING THEORY

This theory is unprovable but **attempts to unite all the forces of nature**, stating that **reality is made of vibrations**—metaphorically called **strings**.

THE PERIODIC TABLE

The periodic table presents all the chemical elements currently known by their atomic number, electron configuration, and chemical properties.

- **Atomic number** is the **number of protons** in the **nucleus**.
- Atomic number tells us the **number of electrons** in an **atom**, as they balance out the number of protons.
- **Atomic mass** is the **number of protons and neutrons** combined.
- Elements are represented by an **atomic symbol**.

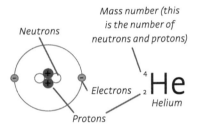

ELECTRON ORBITS

Electrons are arranged in **different orbits** or **shells** around a **nucleus**. **Schrödinger's wave function** tells us that an **electron's position** can be determined by a **probabilistic mathematical equation**. The wave function tells us where an electron *might* be and how electrons **"pack"** around the nucleus. It is helpful to imagine electrons as **"standing waves"** (see Sound and Acoustics, page 36).

THE AUFBAU PRINCIPLE

The different **orbitals** that electrons arrange themselves in are named s, p, d, and f. The aufbau principle tells us **the order in which electrons fill up**, starting with 1s (which can hold only two electrons), followed by 2s, 2p, 3s, 3p, 4s ...

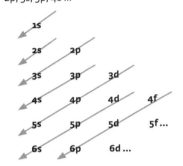

Valence electrons are the outermost electrons associated with an atom, and take part in **chemical reactions**.

READING THE PERIODIC TABLE

- The **vertical groups** in the periodic table tell us **how many valence electrons each element has**. Group 1 has one, group 2 has two, etc.
- The s orbitals are filled first, then p, etc. The periodic table **maps out how orbitals are filled**.
- Moving along the rows from left to right, the **elements are arranged with increasing atomic number**.
- The **noble gases** in group 8 have a **full outer shell of eight electrons**, making them **unreactive**.

Legend:
- Nonmetal
- Alkali metal
- Alkaline earth
- Transition metal
- Metal
- Metalloid
- Halogen
- Noble gas
- Lanthanide
- Actinide

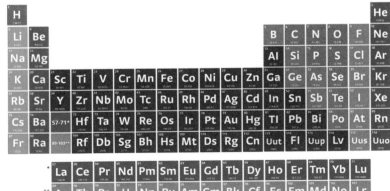

CARBON DATING

Carbon dating is used in many different fields of analysis to determine the age of something.

RADIOACTIVITY occurs when **radiation is spontaneously emitted** by a radioactive nucleus.

ISOTOPES

Some elements will have different numbers of neutrons in their nucleus. **Isotopes** are atoms of an element with different numbers of neutrons, which can make them heavier or lighter. **Atomic mass** is often represented as an average of these masses, which is why the atomic mass of C is in fact 12.011. Carbon can exist as C12, C13, and C14.

HYDROGEN ISOTOPES

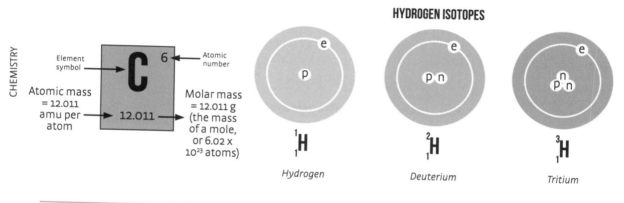

HALF-LIFE

Radioisotopes have their own **rate of decay**. **Half-life** is the time it takes for half of the radioisotopes of a sample to decay.

DECAY CURVE

A decay curve shows the **rate of decay of an isotope**. Taking the example of **tritium** in the graph below.

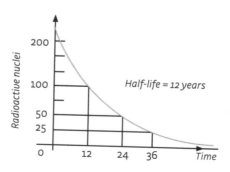

CARBON C14

All **living creatures** contain a tiny amount of **radioisotope C14**. In a living organism, the percentage of C14 stays **constant** as it is continually being replaced by eating and breathing. When an organism dies, the **C14 is no longer replaced** and begins to **decay**. Scientists can work out how long ago it died from the amount of C14 remaining.

INTRAMOLECULAR BONDING

Intramolecular bonds are the forces between atoms that form to make molecules. Bonds occur because of electrons moving from one atom to another or from electron orbitals overlapping.

Electrons are quantum objects that exist as probabilistic superpositions of wave functions.

Lewis structures are **diagrams** used to represent the **bonding in a molecule**. Lone pairs of electrons (unbonded or unshared) in the molecule are shown as dots. The Lewis structure of ammonia (NH_3) is shown here:

$$H-\overset{..}{\underset{|}{N}}-H$$
$$H$$

THE OCTET RULE

In most* cases, an atom is **more stable** with **eight valence electrons** filling its outermost electron shell. When a **covalent** or **ionic bond** occurs, the goal is to achieve a full octet of eight valence electrons.

H_2O (below) shows the outermost single electrons of hydrogen being shared with oxygen's outer six: filling the s orbital of hydrogen with two electrons, and filling the p orbital of oxygen with eight electrons.

Metallic bonding arises from the **electrostatic attractive force** between **positively charged metal ions** and **"dissociated" electrons**, which are free to move under the influence of a **potential difference**.

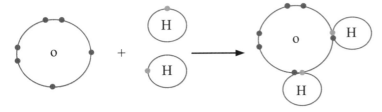

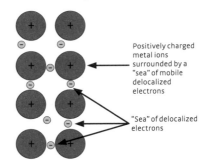

Positively charged metal ions surrounded by a "sea" of mobile delocalized electrons

"Sea" of delocalized electrons

*s orbitals are stable with two electrons; p, d, and f orbitals are stable with eight.

Covalent bonds are formed by **overlapping electron "shells."** This is often described as the sharing of electrons.

Ionic bonding occurs when the **electrostatic difference** between a positively and negatively charged **"ion"** brings them together. In ionic bonding, the **transference of an electron** takes place from one atom to another.

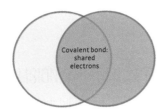

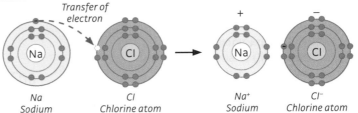

Transfer of electron

Na — Sodium   Cl — Chlorine atom Na^+ — Sodium Cl^- — Chlorine atom

INTERMOLECULAR BONDING

Intermolecular bonds (or forces) are the forces that act between molecules, which include forces of attraction and repulsion. There are three main types:

1. Hydrogen bond
2. Dipole–dipole force
3. Van der Waals force

HYDROGEN BONDING

Hydrogen is a tiny atom that, due to **asymmetrical distribution of electrons around a molecule**, can be slightly positively charged and exert an attractive force on electrons of other polarized molecules. **Hydrogen bonds** are the **reason why so many substances will dissolve in water** and is what makes the solid structure of ice less dense than water, **causing it to float.**

DIPOLE–DIPOLE FORCES

These occur due to an **asymmetrical distribution of electrons in a molecule**, and include forces of attraction and repulsion.

- A molecule of **hydrogen chloride** (HCl) is slightly positively charged around the H atom because the shared electrons in the molecule **bunch around** the Cl atom's **more plentiful protons**.
- The slight positive charge on the H side of HCl attracts the slightly negatively charged Cl side of another HCl atom. This **interaction is a dipole moment.**

VAN DER WAALS FORCE

These forces are distance-dependent interactions between atoms or molecules caused by **polarization in a molecule**. They are a type of dipole moment and are **easy to break apart.**

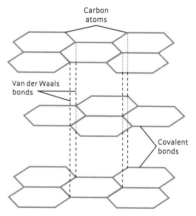

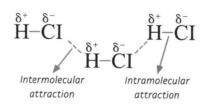

ALLOTROPES

Allotropy or **allotropism** occurs when **elements exist in two or more different structural forms**. This occurs because the atoms of some elements can **bond in different ways**. **Carbon** has many allotropes:

ALLOTROPES OF CARBON

Graphene   *Nanotube*   *Fullerene*   *Diamond*   *Graphite*

CHEMICAL REACTIONS

Without chemical reactions, we would not exist. Chemical reactions take place when bonds between atoms are broken or reformed.

WRITING CHEMICAL EQUATIONS

Chemical equations represent **chemical reactions**. There are four things to consider when writing chemical equations:

1. The **reactants** are what you have at the **beginning of a reaction**; the **products** are what you have **at the end**. Write them as words first: hydrogen + oxygen → water
2. **Replace the words with the formula** for the reactants and the products: $H_2 + O_2 \rightarrow H_2O$
3. **Balance the equation**! There must be an **equal number** of reactant and product atoms on either side of the equations: $2H_2 + O_2 \rightarrow 2H_2O$
4. Include **physical states** of the reactants and products: $2H_2(g) + O_2(g) \rightarrow 2H_2O(l)$

PHYSICAL STATES

- Gas (g)
- Liquid (l)
- Solid (s)
- Aqueous (aq)

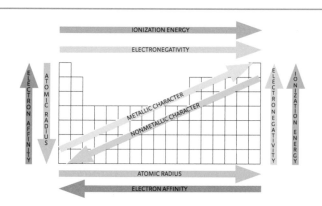

CHEMICAL REACTION INDEX

- **Addition**: two or more atoms or molecules react to form a single molecule.
- **Catalysis**: catalysts increase the rate of reaction but are not used up.
- **Dehydration**: where water is removed.
- **Displacement (or substitution)**: where one atom/molecule takes the place of another.
- **Electrolysis**: a reaction through an electrolyte due to the flow of ions in an electric current; the reaction itself takes place at the cathode and anode.
- **Endothermic**: heat is absorbed.
- **Esterification**: a reaction between an alcohol and an organic acid to form an ester.
- **Exothermic**: heat is given out.
- **Fermentation**: sugars breaking down into alcohol and carbon dioxide.
- **Hydrolysis**: the "decomposition" or breaking down of a compound in water.
- **Ionic association**: oppositely charged ions combine and precipitate.
- **Ionization**: resulting in charged ions.
- **Oxidation**: when oxygen is gained.
- **Polymerization**: small molecules join together to make a long chain molecule.
- **Precipitation**: liquids form a solid product.
- **Redox**: oxidation and reduction take place.
- **Reduction**: when oxygen is lost.
- **Reversible reaction**: reactants can form products and products can reform into reactants.
- **Thermal decomposition**: non-reversible breaking up of a compound by means of heat.
- **Thermal dissociation**: reversible breaking up of a compound through heat.

ORGANIC CHEMISTRY

Organic chemistry is the study, development, and research of the geometric structure, reactivity, and physical and chemical properties of carbon-based (organic) molecules.

LIFE

Carbon chemistry is the **basis of all life** as we know it.

POLYMERS

Carbon readily forms **carbon–carbon bonds** and can form **long molecule chains** in **polymerization reactions**. The long molecules formed are called **polymers** and each unit in a polymer chain is called a **monomer**.

HYDROCARBONS

A carbon atom has four **valence electrons**, making it half full, as ideally it would have eight valence electrons. This is **methane**; it has **one carbon atom** covalently bonded to **four hydrogen atoms**.

```
    H
    |
H — C — H
    |
    H
```

Carbon atoms **can bond together**, instead of a single carbon bonding with four hydrogens. Carbon can form **single**, **double**, or **triple bonds** between other carbon atoms.

NAMING HYDROCARBONS

Hydrocarbons are complicated, so each is **named in such a way that chemists know how many carbon atoms there are and what functional group it is in.**

Number of carbon atoms	Prefix	Molecular formula	Name
1	Meth	CH_4	Methane
2	Eth	C_2H_6	Ethane
3	Prop	C_3H_8	Propane
4	But	C_4H_{10}	Butane
5	Pent	C_5H_{12}	Pentane
6	Hex	C_6H_{14}	Hexane
7	Hept	C_7H_{16}	Heptane
8	Oct	C_8H_{18}	Octane

FUNCTIONAL GROUPS

Functional groups contain similar atoms and **help researchers classify structures** and **predict the behavior of hydrocarbons**. **Alcohols** are a functional group and contain a C–O–H unit.

```
  H   H
  |   |
H—C — C—O—H
  |   |
  H   H
```
Structural formula

C_2H_2OH *Molecular formula*

Other functional groups such as the **alkenes** (with a double bond), **alkynes** (with a triple bond), and **amines** (with a bonded C–NH_2 functional group) are presented here:

Single bond

Double bond

Triple bond

INORGANIC CHEMISTRY

Inorganic chemistry is the study of the structure, properties, and reactions of chemical elements and compounds except for organic compounds (hydrocarbons and their derivatives).

IONIZATION ENERGY

Ionization energy (E^I) is the **minimum amount of energy needed to remove a valence electron from a gaseous atom or molecule**.

VALENCE OR OXIDATION NUMBER

Valence or oxidation number is **the number of electrons available in an atom or molecule that can take part in chemical reactions**.

Group 1: Alkali metals: valence of +1
Group 2: Alkali earth metals: +2
Group 3–6: Transition metals: have multiple oxidation numbers
Group 7: Halogens: –1

ALKALI METALS

- Metal ions with a **+1 charge**.
- **React with water** to **form metal hydroxide ions**, i.e., $M \rightarrow M^+ + e^-$.
- Reactivity increases as you move down group 1, as **valence electrons are farther from the nucleus and easier to remove**.
- Are very **reactive with water and air, emitting light and heat** during a reaction.
- Are very **soft** and **easy to cut**.

ALKALINE EARTH METALS

- Metal ions with a **+2 charge**.
- **React with water** to form metal hydroxide ions—except for **beryllium**, i.e., ionic equation: $Mg \rightarrow Mg^{2+} + 2e^-$.
- Chemical equation: $Ca + 2H_2O(l) \rightarrow Ca(OH)_2 + H_2$.
- Reactivity increases as you move down group 2, as the **valence electrons are farther and farther away from the nucleus**.
- React with **halogen** to form metal **halides**.
- Group 2 **metal sulfates' reactivity** decreases as you move down the group.
- **Hydroxide reactivity increases** moving up the group.
- Are **quite soft**.

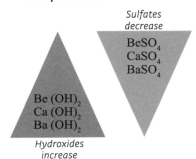

Sulfates decrease
BeSO$_4$
CaSO$_4$
BaSO$_4$

Be(OH)$_2$
Ca(OH)$_2$
Ba(OH)$_2$
Hydroxides increase

TRANSITION METALS

- A single element will have **multiple valency numbers**.
- **Reactivity reduces** across the periodic table.
- Form **brightly colored aqueous solutions**.
- Have **high melting and boiling points**.
- **Solid** at room temperature, **except for mercury** (Hg).
- Are **dense** and **hard**.

General reduced reactivity

HALOGENS

- Nonmetallic elements of group 7.
- Get **less reactive** as you go down the group.
- Halogens **ionize to -1 ion**.
- When a halogen is **ionized, it has the suffix -ide**.
- **Elemental halogens** (substances made of only halogens) such as **chlorine gas** or **fluorine gas** exist as **"diatomic"** (two-atom) molecules.

POWER OF HYDROGEN

Power of hydrogen pH is a measure of H⁺ ions in a solution.

ACIDS

An **acid** is **a substance that contains hydrogen ions** (H⁺). Acids are **proton donors** in chemical reactions:

- Acids form **acidic solutions in water**.
- They are **sources of hydrogen ions**, H⁺.
- For example, **hydrochloric acid** produces hydrogen ions: $HCl(aq) \rightarrow H^+(aq) + Cl^-(aq)$.
- Acidic solutions have **pH values of less than 7**.

ALKALIS

An **alkali** (or base) contains hydroxide ions (OH⁻), will **react with an acid to produce a salt**, and is a **proton acceptor**.

- Alkalis form **alkaline solutions in water**.
- Alkalis are **sources of hydroxide ions**, OH⁻.
- For example, **sodium hydroxide** produces hydroxide ions: $NaOH(aq) \rightarrow Na^+(aq) + OH^-(aq)$.
- Alkaline solutions have **pH values greater than 7**.

ACID REACTIONS

Acid + metal → salt + hydrogen
i.e., hydrochloric acid + magnesium → magnesium chloride + hydrogen
$2HCl(aq) + Mg(s) \rightarrow MgCl_2(aq) + H_2(g)$

Acid + metal oxide → salt + water
i.e., sulfuric acid + copper oxide → copper sulfate + water
$H_2SO_4(aq) + CuO(s) \rightarrow CuSO_4(aq) + H_2O(l)$

Acid + carbonate → salt + water + carbon dioxide
i.e., hydrochloric acid + copper carbonate → copper chloride + water + carbon dioxide
$2HCl(aq) + CuCO_3(s) \rightarrow CuCl_2(aq) + H_2O(l) + CO_2(g)$

ALKALI REACTIONS

i.e., Sodium + water → sodium hydroxide + hydrogen
$2Na(s) + 2H_2O(l) \rightarrow 2NaOH(aq) + H_2(g)$

NEUTRALIZATON REACTIONS

Acid + base = salt + water
$H^+(aq) + OH^-(aq) \rightarrow H_2O(l)$

Acid + metal hydroxide → salt + water
i.e., nitric acid + sodium hydroxide → sodium nitrate + water
$HNO_3(aq) + NaOH(s) \rightarrow 2NaNO_3(aq) + H_2O(l)$

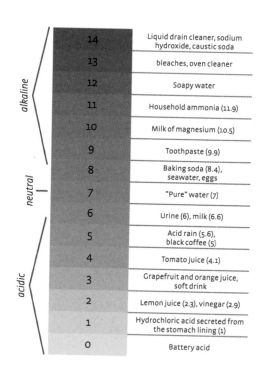

HYDROGEN BONDING AND WATER

Water is the reason why life exists on Earth; it makes up 71 percent of the Earth's surface and living organisms contain between 60 and 90 percent water.

POLARITY

This is caused by the **asymmetrical electron distribution** around a molecule and a property called **electronegativity**. For a molecule to be polar there has to be a **"dipole moment,"** which is **a separation of the charge around the molecule between a positive side δ+ and a negative side δ-**.

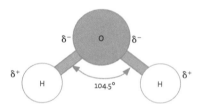

ELECTRONEGATIVITY

Electronegativity is a **measure of the degree to which electrons in a bond are attracted to an atom**.

- The **Pauling scale** is used to measure electronegativity.
- Electronegativity **increases from left to right in the periodic table**.
- Moving down the rows of the periodic table, electronegativity **decreases due to shielding caused by the extra layers of electron orbitals that shield the positive-charge influence of the protons**.
- **Fluorine** is the **most electronegative** of all the elements.

- **Ionic bonds** are generally more electronegative than covalent bonds, with the exception of **asymmetric charge distribution** occurring in **some hydrogen-bonded molecules** such as water.

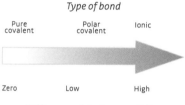

THE POLARITY OF WATER

Solvents that contain polar molecules (such as water) are **very good at dissolving solutes. Water dissolves more compounds than any other solvent.**

COSMIC WATER

- In 2011 astronomers discovered a **cloud of water ice surrounding a black hole** that contained 140 trillion times more water than there is on Earth.
- **Europa**, one of **Jupiter's icy moons**, is covered in a crust of ice, specifically **salt water ice**.
- **Enceladus**, one of **Saturn's moons**, also has a water ice crust and ocean below.
- **Comets** are made of rock and water.

GEOMETRY OF WATER MOLECULES

Oxygen has six **valence electrons** and needs two more electrons from **hydrogen atoms** to complete its octet; **covalently bonding with hydrogen makes** that possible. As the oxygen atoms contain more **protons** and the hydrogen atom has only one, electrons in the molecule tend to **gather around the oxygen atom, causing a dipole moment**.

USEFUL TERMINOLOGY

- **Solvent**: a substance that will dissolve a substance as a solid, liquid, or gas; solvents can be aqueous (contain water) or nonaqueous (do not contain water).
- **Solute**: a substance that has been dissolved by a solvent.
- **Solution**: a solute disolved in solution.

COHESIVE FORCES

Polar molecules in solution like to **arrange themselves according to their charge**; the **dipole moment** within the molecule **causes it to stick to itself**. This is why **water droplets** form **spherical-shaped beads**.

STATES OF MATTER

Matter is distinguished by its properties. Some properties are sensory, some are chemical, and some are physical. A change in state is a physical property.

PHYSICAL PROPERTIES INCLUDE:
- Density
- Molecular geometry
- Specific gravity
- Odor
- Color
- Accidental properties: arising from other phenomena include **texture, shape, volume,** and other sensory properties important to material scientists and designers.

Adding heat to a system will give molecules or atoms more **kinetic energy**, causing them to jiggle around with increasing **kinetic speed**. When you apply more heat or energy to a system, **you can change the "state" of a substance.**

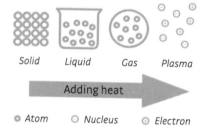

PHASE CHANGE

The **thermal energy** of a system provides the molecules of a system with enough energy to **vibrate**. With that, they **can overcome intermolecular or intramolecular forces that hold them in place as a solid**—resulting in **different states of matter**.

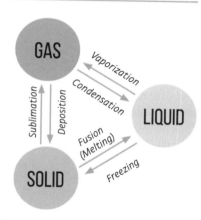

BROWNIAN MOTION

Atoms or molecules **in a fluid** (liquid or gas) will **move in a random, erratic path** and continuously **bump into other molecules**. This was discovered by Scottish botanist **Robert Brown** in 1827.

RANDOM MOTION OF A PARTICLE IN A FLUID

Brownian motion will result in the **diffusion of particles in a fluid** where, due to random bombardments, **particles end up being equally dispersed throughout the fluid**—reaching an **equilibrium**.

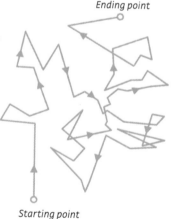

OTHER STATES

Plasma is a state of matter cooled to near absolute zero (0 K, −459.67 °F) causing matter to act like a single quantum mechanical entity whose behavior can be described by the wave function.

PLASMA

Plasmas contain a lot of energy. In a plasma, there is **so much energy that electrons are stripped from positively charged nuclei, forming a soup of freely flowing ions**. Plasma is the **most abundant state of matter in the universe**. Stars and **stellar supernovae** are in the plasma state.

CHIRALITY

Chirality is a geometric property that some molecules and ions possess. A chiral molecule is a non-superimposable mirror image of itself. Chiral molecules have the same chemical formula but different chemical properties.

Isomers: compounds with the same chemical but **different atomic arrangements**, which result in different properties.

Stereoisomers: molecules that have the same chemical formula but **different spatial arrangement** of their atoms.

Diastereomers: stereoisomers that are **non-superimposable** on each other—or are **mirror images** of each other.

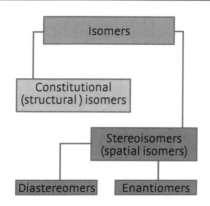

Chiral compounds
- Have the same chemical formula;
- Have different geometric formulae;
- Have different chemical properties.

Cis–trans isomers are types of stereoisomers:

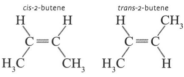

cis-2-butene — Methyl groups on opposite side of double bond

trans-2-butene — Methyl groups on same side of double bond

S AND R ENANTIOMERS

Asymmetries surrounding a carbon-atom-centered molecule cause chirality in organic molecules. Enantiomers are **pairs of molecules that are mirror images of each other**.

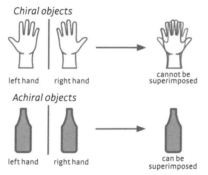

Chiral objects — left hand / right hand — cannot be superimposed

Achiral objects — left hand / right hand — can be superimposed

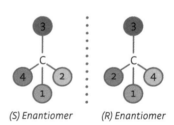

(S) Enantiomer   (R) Enantiomer

SUGAR ENANTIOMERS

In biology, enantiomers often react with organisms differently, because molecules frequently **bind to receptors in the body that are also chiral**. The enantiomers of **glucose** (D-glucose and L-glucose) react in the body differently. Our bodies can use **D-glucose** for energy but not L-glucose. **L-glucose** can be made only in a laboratory and is not found in nature.

DNA AND CHIRALITY

The **double helix** of DNA twists in a clockwise direction. The molecules that make DNA have **chiral centers** themselves, causing the intramolecular bonds in the DNA to **spiral** in restricted directions. DNA is a **"right-handed" helix**.

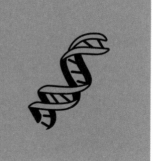

MACROMOLECULES

Macromolecules are giant molecules composed of thousands of atoms or more. Biopolymers and large non-polymeric molecules are some of the most common.

MOLECULAR STRUCTURES

Understanding the **three-dimensional shape, geometry,** or **configuration** of the **smaller subunits of molecules** helps us understand their chemistry and why they are structured the way they are. **The ways in which electron orbitals interact with each other is what defines the geometry of all molecules.**

MOLECULAR GEOMETRIES

Small molecules have different geometries:

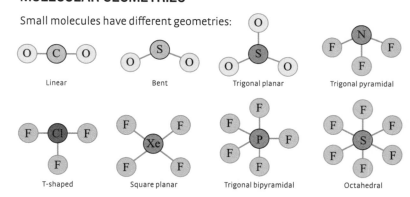

AMINO ACIDS

There are **twenty amino acids necessary for making proteins**. Sometimes referred to as **"the R-group,"** different proteins will have different properties depending on what amino acids they contain. The diagram shows the basic structure of an amino acid. A molecule referred to as the R-group is what makes each amino acid different.

AMINO ACID STRUCTURE

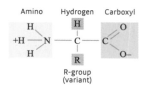

PROTEINS

Proteins are made of **chains of amino acids**. Each amino acid in a protein starts as a single small molecule. **Polypeptides** are chains of amino acids.
- **Primary structure**: polypeptide chains.
- **Secondary structure**: when a **polypeptide chain** reaches a certain size, it folds back or twists around itself, forming **alpha helices** and **beta sheets**; this twisting or folding arises from intermolecular bonds interacting with each other.
- **Tertiary structure**: these arise from **multiple secondary structures packing together**; intermolecular bonds dictate how they are structured together.
- **Quaternary structure**: some proteins are very **complex** and are made up of two or more tertiary structures folding around themselves; **hemoglobin** is a quarternary structure.

PROTEIN FOLDING

The folding in a protein is very **difficult to predict**; there are a **vast number** of ways that intermolecular bonds formed by the different amino acids interact. From **hydrogen bonds** to **hydrophobic and hydrophilic centers**, all these forces influence how a protein is formed. A **disulfide bridge** occurs within a polypeptide chain and **can stabilize tertiary structures**.

POLYMERS

Polymers are large molecules made from repeated monomer subunits. Polymerization is the chemical joining of monomers, resulting in polymers.

Addition polymerization: monomers join together **like beads on a necklace** with the help of a catalyst.

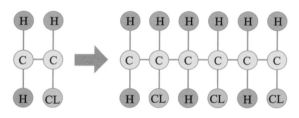

Condensation polymerization: monomers polymerize and create **water, carbon dioxide**, or **ammonia** in the process.

NATURALLY OCCURRING POLYMERS
- **Polysaccharides**—carbohydrates such as **starch, cellulose, glycogen**, and **pectin**.
- **Cellulose**—a **polysaccharide** found in **vegetables, plants**, and **trees**, made from **sugar molecules**; form **long, flexible fibers** used in **textiles**.
- **Pectin**—a **polysaccharide** with a **jelly-like texture**.
- **Silk**—produced by **silkworms** (which are caterpillars) when they spin cocoons.
- **Spider silk**—a very **robust** polymer made from **proteins** by **spiders**.
- **Wool**—made from a protein called **keratin**. Sheep grow wool to keep them warm; it can safely be sheared off (shaved) and used in **textiles**.
- **DNA**—made from sugar and molecules called **nucleotides**.
- **Proteins**—made from **amino acids monomers**.
- **Collagen**—a **fibrous** polymer present in the **muscle** and **connective tissue** of **animals, birds**, and **fish**.

PLASTICS

Plastics can be **heated, melted, formed**, and **reformed**. Many plastics are derived from **crude oil**. Plastic is a major cause of **pollution**. It breaks down into **microplastics**, which are **toxic to all life**. Cutting down **single-use plastic** is helpful; however, some single-use plastics are essential in **medical contexts**. All plastics CAN be **recycled** and **do not biodegrade**.

TYPES OF PLASTIC
- Polyethylene terephthalate (PETE or PET)
- High-density polyethylene (HDPE)
- Polyvinyl chloride (PVC)
- Low-density polyethylene (LDPE)
- Polypropylene (PP)
- Polystyrene or Styrofoam (PS)

BIOPLASTIC

Plastics can be made from **biological sources**, i.e., **sugars, starches from wood chips**, and **food waste**. Bioplastics are **biodegradable**, but some can be broken down only by industrial composting, which requires temperatures of 160°F.

TYPES OF BIOPLASTIC
- Protein-based
- Starch-based
- Cellulose-based
- Polylactic acid (PLA)
- Casein (milk proteins)
- Lipid-derived polymers

HYDROPHILIC AND HYDROPHOBIC

Hydrophilic means "water loving" and hydrophobic means "water fearing."

Hydrophilic
+ low adhesion
+ easy water removal
− fast ice growth

Hydrophilic surface

Hydrophobic
+ low adhesion
− slow ice growth
− pinning of drops

Hydrophobic surface

- **Hydrophilic** substances **dissolve in water**. They include **ionic compounds** such as **salts** or **polar molecules** such as **alcohols**.
- **Hydrophobic** substances **do not dissolve in water**. **Oils** and **fats** will **repel water** and are **nonpolar**.

WATER AND OIL

Droplets of **oil** will **float on water** and will **not mix**. They are referred to as **immiscible**.
- **Polar solvents**: have an **asymmetrical electron distribution**.
- **Nonpolar solvents**: electrons are **evenly distributed** and are **geometrically symmetrical**.

CELL MEMBRANE

The cell membrane (also called the **lipid bilayer**) exists due to its **hydrophilic hydrophobic properties**. The cell membrane is made of two layers, with **polar outer-facing heads** that enable it to exist in the environment of our bodies, while **hydrophobic tails** face in.

INSULIN

Insulin is a protein that **controls blood sugar levels**. It **helps cells absorb (metabolize) glucose (energy)**. When cells have enough glucose, the **liver** stores it as **glycogen** (which prevents blood glucose levels dropping too low). **Insulin** cannot directly access its target cells because it is **hydrophilic** and can't pass through the **hydrophobic inner bilayer**. It relies on **signaling** in order to function. Without insulin, **receptors** cannot tell how much glucose is in the blood and it **cannot be metabolized**.

AMINO ACIDS

Twenty amino acids are needed to form all the **proteins** that **constitute** and **sustain** our bodies (see illustration below). The predominantly **aqueous environment** of the body requires hydrophobic and hydrophilic amino acids.

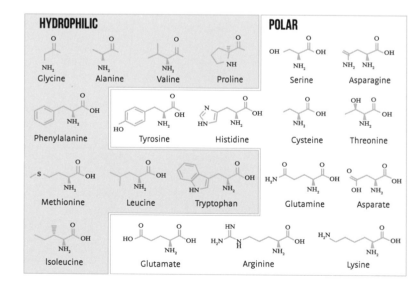

PROTEIN CRYSTALLOGRAPHY

Protein crystallography is used to determine the atomic structure of molecules in a crystallized substance. It is widely used in chemistry, physics, and the analysis of biological molecules such as protein and DNA.

DOROTHY HODGKIN

British structural biologist **Dorothy Hodgkin** (1910–94) researched **protein structures** using **X-ray crystallography**. She was awarded the **1964 Nobel Prize in Chemistry** for discovering the structure of **vitamin B12**. She developed innovative techniques for **imaging the three-dimensional structures of molecules** and helped to resolve the structure of **penicillin, insulin,** and **steroids**.

THE COMPLEX STRUCTURE OF VITAMIN B12

Vitamins are **molecules** needed by the body, most of which **the body cannot produce itself**. B12 is a water-soluble molecule and is necessary for the **maintenance and production of nerve cells, blood cells, and DNA**.

CRYSTALLIZING A PROTEIN

Proteins have **very complex structures**; discovering **their structure helps us to understand their function in the body**. **Understanding how insulin is formed** has helped us to **produce it in the lab** and save the lives of people with **type 1 diabetes**. This technique is helpful in understanding the structure of complex biological molecules.

X-RAY DIFFRACTION

This is a technique used to determine the **structure of crystals**. **X-rays** have a very small **wavelength** and can be made to pass through a crystal. As they pass through, they create a **diffraction pattern**. The patterns tell us **how the atoms are arranged in a lattice**, the **distance between atoms**, and the **scale of the lattice**.

CRYSTALS

Crystals are solids. They form as **regular structures** with flat surfaces and straight edges. They consist of millions of small particles arranged in a **repeating pattern** called a **lattice**. The regularity of crystals makes it easy to **geometrically analyze their molecular structure**.

COMMON CRYSTAL GEOMETRY

Crystals can form with many different geometries, but the four most common ones are **cubic, hexagonal, monoclinic,** and **rhombic**.

Cubic   *Hexagonal*   *Monoclinic*   *Rhombic*

DNA AND PHOTO 51

X-ray crystallography is a way of studying the internal structure of materials from how they interact with short-wavelength X-rays. In the early 1950s it revealed the structure of the DNA molecule that is key to life itself.

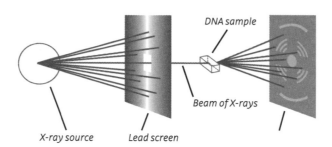

THE DOUBLE HELIX

Although biologists had **discovered DNA** many years before and recognized its role in **storing the genetic information** used by cells to replicate and make other chemicals, it was only in the early 1950s that they began to understand how it worked.

Pioneering X-ray work by **Rosalind Franklin** and colleagues at King's College, London, began to pin down the presence of key chemical structures within the enormous molecule. In 1953 a crucial image known as **"Photo 51"** inspired Cambridge biologists **Francis Crick and James Watson** to build a model in which **pairs of chemical units** called **bases** or **nucleotides** were linked together to form the rungs of a twisting-ladder shape called a double helix.

Unfortunately the DNA story became mired in **controversy** after Watson's best-selling account of the discovery attempted to minimize Franklin's important role, often with **outright sexism**. Furthermore, it seems the Cambridge pair gained **access to crucial research without her knowledge**, in part through their mutual colleague Maurice Wilkins.

The true story and the motivations of those involved will probably never be known, and the opportunity for a resolution was curtailed by Franklin's death from ovarian cancer in 1958.

Four years later, **Watson, Crick, and Wilkins were awarded the Nobel Prize for "their" discovery**, but in recent years Franklin has gone from the **"forgotten woman of DNA"** to an **iconic figure**.

FROM PHOTO 51 TO THE DOUBLE HELIX

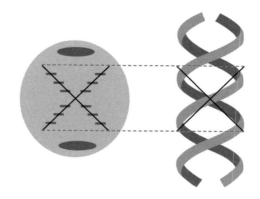

THE CENTRAL DOGMA OF BIOLOGY

The central dogma of biology is: DNA makes RNA makes protein makes DNA. DNA is tightly packed into chromosomes, which are found in the cell nucleus.

DNA

- Made of **nucleotides**: **adenine** (A), **guanine** (G), **thymine** (T), and **cytosine** (C).
- A, G, T, and C **bond** between two **sugar-phosphate chains** that **twist like a helical ladder**.

BASE PAIRS

A bonds with T; C with G.
The order (or sequence) of these pairs allows DNA to **encode information**.

 =

Purines = Pyrimidines

- There are approximately **six billion base pairs in every cell**.
- DNA: **Deoxyribonucleic acid**—a **double helix**.
- RNA: **ribonucleic acid**—a **single helix**, contains **uracil** instead of **thymine**.

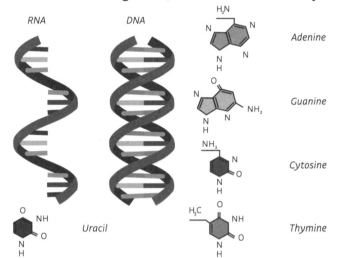

REPLICATION

Cells **copy DNA trillions of times** in a lifetime, using half the DNA as a **template**. Enzymes are proteins that **control replication**:

- **Helicase** unzips DNA, to create a template.
- **RNA primase** begins the process.
- **DNA polymerase** adds complementary nucleotides.
- **DNA ligase** ends the process.

Errors occur once in every ten billion nucleotides. DNA polymerase checks DNA to **minimize error**.

TRANSCRIPTION

Genes are **expressed through** the **production of proteins**. Segments of **DNA** are copied to form **mRNA** (instructions for making proteins).

- **RNA polymerase unzips the double helix**—copies the sequence.
- RNA reaches a **terminating signal and stops**.
- **Messenger RNA (mRNA)** is formed.
- mRNA **leaves the nucleus to make protein**.

TRANSLATION

Ribosomes make **proteins** in the **cytoplasm**.

- mRNA is fed into the **endoplasmic reticulum**.
- Ribosomes contain **protein** and **ribosomal RNA (rRNA)**.
- rRNA reads the mRNA **three nucleotides at a time**.
- tRNA has an **amino acid at one end** and bases called **anticodons at the other** that match mRNA.
- **Amino acids** carried by tRNA **bond to form a polypeptide chain**—the beginning of a protein.

CELLS

Cells are the most basic component of an organism.

PROKARYOTIC CELL

A prokaryote is a **single-celled organism**. It has **"flagella" to move with**, and **pili to sense surroundings**.

EUKARYOTIC CELL

These are found in **multi-celled organisms**, and have a **more complex interior**.

WHAT'S IN A CELL?

- **Cell membrane**—**surrounds** and **protects** the cell, permeated by **nutrients** and **waste**.
- **Nucleus** (in eukaryotes)—**controls** the cell, **contains DNA**.
- **Cytoplasm**—**chemical reactions** happen in this **complex material**.
- **Mitochondria**—**metabolism** (**respiration**) takes place.
- **Ribosomes**—found in the **endoplasmic reticulum**; make **proteins**.

SPECIFIC TO PLANT CELLS

- **Cell wall**—**rigid outer structure**.
- **Vacuole**—contains cells that contain **sap**.
- **Chloroplasts**—where **photosynthesis** happens.

CELL ANATOMY

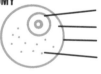
Nucleus
Cytoplasm
Cell membrane
Mitochondria

Cell wall   Chloroplasts

CELLULAR RESPIRATION

Cells get **energy** through chemical reactions called **respiration**, where **carbohydrates are broken down in the cytoplasm**.

- **Endothermic**: chemical bonds breaking **require energy**.
- **Exothermic**: chemical bonds breaking **release energy**.

Energy released in **exothermic reactions** combines with **phosphate ions** to make a molecule called **adenosine triphosphate** (ATP). **Oxygen** (O_2) **accepts electrons** while **carbon dioxide** (CO_2) **is released**.

In **photosynthesizing** cells, CO_2 is used to form **carbohydrates**. O_2 is released as **waste**.

GLYCOLYSIS

This is a simplified description of cell respiration:

- **Glucose** is broken down into **pyruvic acid** (part of the **"Krebs cycle"**).
- Pyruvic acid is broken down to release more **energy**—to create **nicotinamide adenine dinucleotide** (NADH).
- This **transports electrons** along a **chain of enzymes and molecules** called **"cytochromes"**; electrons are the source of the energy.
- Electrons **repel protons** (H^+ ions) through a **membrane** where they **make ATP**—this is **chemiosmosis**.
- **Glycolysis** does not use **oxygen**—this process is **anaerobic**.
- Some **bacteria** and **yeasts** use only glycolysis for energy.

THE KREBS CYCLE (CITRIC ACID CYCLE)

Molecules of **pyruvic acid** form **ATP, CO_2, NADH**, and **flavin adenine dinucleotide** ($FADH_2$)—a **redox-active molecule**.

THE MICROSCOPE

Without the development of the microscope, we would not have a vaccine for polio or be able to make microchips.

OPTICAL MICROSCOPE

These use visible light and lenses to magnify organisms. Dutch spectacle maker **Zacharias Janssen** is said to have designed the **first microscope** in 1595, which **magnified up to nine times**.

ROBERT HOOKE'S *MICROGRAPHIA*

English scientist **Robert Hooke** (1635–1703) made a microscope and published a book called *Micrographia* in 1665. It was the first ever book published with **illustrated observations of microscopic objects**.

ANTON VON LEEUWENHOEK

Dutch polymath **Von Leeuwenhoek** developed the microscope further and could **magnify objects by 270 times**.

THE ULTRAMICROSCOPE

Austrian chemist **Richard Zsigmondy** invented the **ultramicroscope**. It **focused a light beam** through a **microscopically dispersed suspension of particles** called a **colloid** with particles intended for study in the colloid. It **magnified up to 100,000 times**. Zsigmondy received the **Nobel Prize** in 1926.

PHASE-CONTRAST MICROSCOPE

Frits Zernike designed the phase-contrast microscope in 1932, which made it possible to **view transparent materials and magnified to near atomic resolution**.

ELECTRON MICROSCOPE

First invented in 1931, the electron microscope uses **beams of electrons instead of light** to **magnify objects up to 10 million times**. As it uses electrons, it **can see things smaller than a wavelength of light**. Modern electron microscopes need a **computer** and **software**.

ELECTRON TUNNELING MICROSCOPE (ETM)

Invented at **IBM** in 1982, the ETM operates on the **atomic scale**, revealing the surfaces of materials at the scale of the atoms. It uses **quantum tunneling** to do so. A **computer** and **software** are also needed.

Basic design of the optical microscope from the 1800s

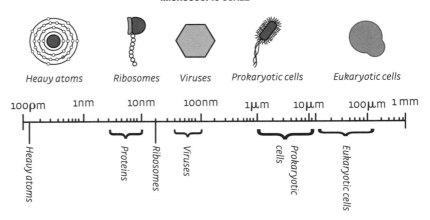

MICROSCOPIC SCALE

MICROBIOLOGY

Microbiology is the study of microorganisms. These include bacteria, viruses, archaea, fungi, and protozoa. It is important to the research fields of medicine, biochemistry, physiology, cell biology, ecology, evolution, and biomedical engineering.

Microorganisms help us **digest food, make cheese, and get colds**. They are **essential to our existence** and to ecosystems everywhere in the world. They are abundant and important, yet some cause deadly diseases.

THE FIVE TYPES OF MICROORGANISM:

Bacteria

Viruses

Algae

Fungi: yeast and mold

Protozoa

ARCHAEA, BACTERIA, AND PROTOZOA

These were the **first organisms to evolve**. Many are found in **extreme environments**; some are found **in and on the body** and **protect us**.

MICROBIOLOGY AND MEDICINE BETWEEN 1796 AND 1929

1715	**Lady Montagu** introduced **Turkish variolation** to her children, which protected them from **smallpox**
1796	**Edward Jenner** discovered a **vaccination for smallpox**
1838	**Matthias Jakob Shleiden** discovered **plants are made of cells**
1840s	**Ignaz Semmelweis** discovered that **hand washing reduces the spread of disease**; few people believed him
1850s	**Rudolf Virchow** discovered that **cells come from cells**—that they **reproduce to make new cells**
1854	**John Smith** identified that a **cholera** outbreak in **Soho, London**, was linked to a **communal water pump**
1864	**Louis Pasteur** developed **pasteurization** and proposed the **germ theory**: **infectious diseases** are caused by germs and **spread by people**
1876	**Robert Koch** founded **bacteriology**; understood that **different kinds of microbes cause different kinds of diseases**
1882	**Franny Hesse** invented **agar plate**
1860s	**Joseph Lister** developed **antiseptic chemicals** that reduce the spreading of disease
1905	**Florence Nightingale** discovered the importance of **cleanliness in patient care** and reducing infectious diseases
1928	**Alexander Fleming** discovered **penicillin**—**antibiotics** are discovered

PASTEURIZATION

French chemist Louis Pasteur (1822–95) discovered how vaccination could prevent disease and how microbial fermentation worked, and invented pasteurization. His breakthroughs have saved countless lives.

Scientific discoveries are **cumulative** and **collaborative**. Pasteur could develop his work only thanks to the work of others before him.

FERMENTATION

Pasteur's early work showed that **fermentation** results from the **activity of living microorganisms**. During fermentation, **sugars** are converted into **ethanol** and **carbon dioxide**, making wine and beer.

THE GERM THEORY OF DISEASE

Pasteur proposed the **germ theory of disease**: infectious diseases are caused by germs and spread by people.

The **medical establishment** was **reluctant to accept Pasteur's germ theory**; they looked down on him because he was a **chemist**. Despite this, Pasteur developed techniques in **vaccination** and contributed to our knowledge of **immunology**.

VACCINATION

Pasteur developed vaccinations for **rabies**, **anthrax**, and **chicken cholera**.

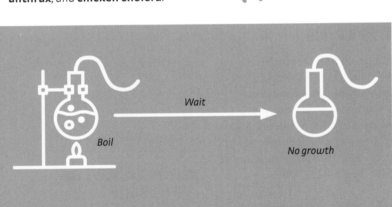

PASTEURIZATION

Pasteur demonstrated in an experiment that **boiling milk** to a temperature of 140–212 °F and **leaving it to cool and settle** in a **swan-neck flask** did not result in **microbial growth** later on. The contents of the flask would not spoil. **If the flask was broken, microbial growth would occur.**

The "swan-neck" bend in the flask that he used stopped microbes from entering the flask. This is why **toilets** have an **s-bend**.

Pasteurization is still used today to **help food products keep for longer**.

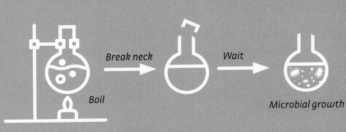

VACCINATION

A vaccination is a medicine that trains your immune system to be able to identify and attack a pathogen by exposing it to a weaker version of said pathogen. This exposure stimulates natural immunity.

SMALLPOX

A **highly contagious virus** that causes **fever** and **clusters of pustules** in hosts, smallpox has killed billions of people, but was **effectively eradicated through vaccination** by 1979.

VARIOLATION

In 1022 a **Buddhist nun from Seishun, China**, was known for **grinding up smallpox scabs** and **blowing them into the nostrils of healthy people**. Many of these individuals developed **immunity** from smallpox. A similar practice spread to **Turkey**.

VACCINATION EXPERIMENT

Edward Jenner (1749–1823) noticed that **milkmaids with cowpox never caught smallpox**—cowpox and smallpox are part of the **poxvirus** family. Jenner injected people not infected with smallpox with cowpox, waited a few months, and then injected them with smallpox. The result was that they were **not infected**.

INOCULATION

- **Cowpox** is **less aggressive in humans than in cows**.
- Introducing humans to cowpox introduced the **immune system** to a less aggressive version of smallpox.
- Patients later introduced to smallpox have immune systems that **recognize the virus** and **know how to attack it**.

MMR (MEASLES, MUMPS, AND RUBELLA) VACCINE

The MMR vaccine protects children from three terrible **viruses: measles, mumps**, and **rubella**. There is **no credible connection between the MMR vaccine and autism**—claims connecting the two are demonstrably false. Autism is NOT a disease; it is like eye and skin color—**part of human diversity**. It **should not be stigmatized**.

HPV VACCINE

The HPV vaccine protects against the **HPV virus**, which is a **known cause of cervical cancer**.

EBOLA

The ebola virus causes **death from bleeding**. Ebola has **evolved to survive in humans**; when humans are infected it has particularly harsh effects. **Vaccinations are under development.**

BACTERIOLOGY

Bacteria are everywhere; some protect us and some cause disease. Bacteria in the blood can be deadly.

Bacteria are **some of the oldest organisms on the planet**; they have existed for over three billion years and make up the vast majority of **prokaryotes** on Earth.

HOW BACTERIA WORK

Bacteria **do not reproduce sexually**; they engage in **horizontal gene transfer**, which is how they can quickly evolve **resistance to antibiotics**. Many bacteria are **parasitic**. They are classified in three main groups:

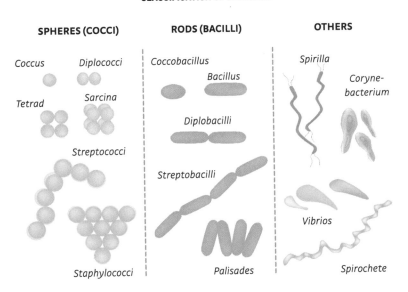

CLASSIFICATION OF BACTERIA

SPHERES (COCCI): Coccus, Diplococci, Tetrad, Sarcina, Streptococci, Staphylococci

RODS (BACILLI): Coccobacillus, Bacillus, Diplobacilli, Streptobacilli, Palisades

OTHERS: Spirilla, Corynebacterium, Vibrios, Spirochete

BACTERIAL VIRULENCE

This denotes the characteristic that increases how pathogenic a bacterium is.

BACTERIAL RESISTANCE

Methicillin-resistant staphylococcus aureus (MRSA) is a **"superbug"** bacteria that has evolved **resistance** to many **antibiotics**. **Infections** with MRSA are **harder to treat** than other bacterial infections.

LIST OF BACTERIAL INFECTIONS

- Bacterial meningitis
- Pneumonia
- Tuberculosis
- Upper respiratory tract infection
- Gastritis
- Food poisoning
- Eye infections

ANTISEPTIC HISTORY

Ignaz Semmelweis (1818–65) was a **Hungarian doctor**, and the **first to introduce antiseptic procedures** after noticing that **washing hands between treating patients saved lives** by **preventing the spread of infection**. The **medical community** at the time were outraged and **refused to wash their hands**—the thought that doctors could spread disease offended them—but Semmelweis was right.

British surgeon **Joseph Lister** (1827–1912) developed **antiseptic surgery**, enabling doctors to carry out **essential surgery** while **lowering the risk of infection and death**.

English social reformer, nurse, and statistician **Florence Nightingale** (1820–1910) **saved the lives of wounded soldiers** who were **previously dying of infection rather than from their wounds**. Her approach to nursing proved the importance of **cleanliness**.

VIROLOGY

In many ways, viruses are not alive at all. They can be described as intracellular (within cell) parasites, and cannot replicate or spread without a host.

VIRUSES

Viruses include the **common cold**, **hepatitis**, **tuberculosis**, **H1N1** (swine flu/bird flu), **rabies**, **HIV**, **ebola**, **HPV** (herpes), **influenza**, **measles**, and **chicken pox**.

- Viruses **consist of a piece of DNA (or RNA) enveloped in protein**.
- Viruses **hijack host DNA and cellular mechanisms** to **reproduce** themselves.
- Viruses **can infect all organisms**: bacteria, bugs, and plants.

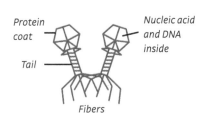

ANATOMY AND SIMPLIFIED LIFE CYCLE OF VIRUSES

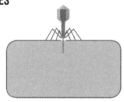

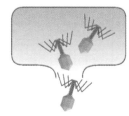

VARIETY

The size of viruses varies from being on the micrometer scale to the nanometer scale.

VIRAL VIRULENCE

Virulence is a characteristic of a virus that increases how **pathogenic** it is. It can decrease or increase over time.

HIV—THE VIRUS THAT CAUSES AIDS

Since the 1930s, about 70 million people have been **infected** and 35 million have **died** of AIDS across the world. The history of HIV is one of great loss and **injustice** for those neglected due to the stigma of AIDS. HIV is **no longer a death sentence** if you can access the correct medicines.

French virologist Françoise Barré-Sinoussi was co-awarded the **2008 Nobel Prize in Physiology and Medicine** with **Luc Montagnier** for discovering both HIV, and, with **Harald zur Hausen**, the **HPV viral cause of cervical cancer**.

RETROVIRUS

A retrovirus **makes use of host RNA to procreate**. The process involves five distinct steps:
1. **Attachment**: bacteriophage virus attaches to host cell.
2. **Penetration**: injects virus DNA/RNA into the cell.
3. **Biosynthesis**: virus DNA/RNA replicates and forms protein.
4. **Maturation**: virus proteins assemble to form new viruses.
5. **Lysis**: new viruses released from cell.

Host–virus DNA interaction can result in chunks of **viral DNA/RNA being interlocked with our own**. If this doesn't **mutate or damage our DNA**, it **stays there**. If this happens in **germ cells** (sperm or eggs), **traces of viral DNA are inherited with no ill effect**.

EXTREMOPHILES

Extremophiles live in what we would consider to be extreme environments. But did extremophiles adapt to extreme conditions or did they evolve from a more primitive life form?

TYPES OF EXTREMOPHILE
Radiation-resistant extremophiles
Acidophile: adapted to acidity pH: 1 to 5
Alkaliphile: adapted to basic conditions pH: 9 to 14
Thermophile: withstands heat
Psychrophile: withstands extreme cold
Thermoacidophile: adapted to hot, acidic conditions
Xerophile: withstands extreme dryness
Barophile (Piezophile): withstands extreme pressure
Halophile: adapted to very salty conditions

HYDROTHERMAL VENTS

Hydrothermal vents **heat the water** to 750 °F and emit **hydrogen sulfide gas** that would **kill other organisms**. These vents support entire ecosystems: an archaea called *Picrophilus torridus*, **Antarctic krill**, and so-called **Pompeii worms**.

CRYPTOENDOLITHS

These **live within porous rocks beneath the Earth's surface**. They have been found in deep rocks in **Antarctica** surviving **independently of the Sun** and have a **biomass greater than that of all the organisms of the Earth's oceans**.

ANCIENT BACTERIA

Ancient **bacterial spores** discovered in a **salt crystal 1,850 feet underground** in the year 2000 were **brought back to life after 250 million years**.

TARDIGRADES

Known as **water bears** or **moss piglets**, these **near-microscopic invertebrates can survive in the vacuum of space, withstand pressures 6,000 times that of the atmosphere**, and **can go into stasis** (suspended animation).

SPACE

Bacteria **cannot survive the vacuum of space**, but hiding in a cavity could protect them from radiation. *Clostridium botulinum* forms **spores that can survive in space**.

ASTROBIOLOGY

Astrobiology brings **geochemistry, biochemistry, astronomy, geophysics**, and **ecology** together to explore the **origins of life**, **early evolution**, and **future life**. Astrobiologists ask if extremophiles could **evolve elsewhere in the solar system and universe**.

ENCELADUS AND EUROPA

Saturn's moon Enceladus and **Jupiter's moon Europa** are **covered in liquid water beneath an icy crust**—conditions similar to some of the Earth's oceans.

MARS

Life cannot survive the radiation or conditions on Mars, but **may have existed there in the past**.

BIOMATERIALS

Biomaterials interact with or are synthesized by biological systems. They are used in pharmaceuticals, medicine, and, increasingly, in architecture, design, and textiles.

BIOMINERALS

These are minerals produced by biological systems—e.g., **skeletons**, **feathers**, **tusks**, or **shells**. Organisms produce biominerals to **protect and reinforce** themselves, and to **sense their environment**. Many organisms form biominerals:
- **Silicates**: sponges, algae, and diatoms.
- **Carbonates**: invertebrate shells.
- **Calcium phosphates and carbonates**: in vertebrates.
- **Copper and iron**: used by some bacteria.

SHELLS AND REEFS

Coral reefs and shells **synthesize robust structures** from **dissolved carbon**, turning it into **calcium carbonate** ($CaCO_3$). **Ocean acidification** poses a great risk to these organisms, as acid dissolves $CaCO_3$.

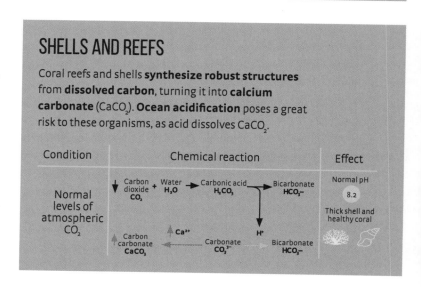

HYDROXYAPATITE

This **complex crystal** ($Ca_5(PO_4)_3(OH)$, or $Ca_{10}(PO_4)_6(OH)_2$) is important in **bone structure**. It **strengthens in the direction of stress**. Babies are not born with solid **kneecaps**; when they learn to walk, their kneecap cartilage begins to **calcify** (form calcium-containing crystals) and **solidify**—thanks to the material properties of hydroxyapatite.

JELLYFISH

Jellyfish orientate themselves by **sensing the Earth's gravitational field** through **particles of $CaSO_4$** embedded in **proteins** in a special membrane.

MAGNETOTACTIC BACTERIA

This diverse group of **water-dwelling bacteria** navigate along the **magnetic field lines** of the Earth. **Iron crystals** of Fe_3O_4 enable this—iron (Fe) is **ferromagnetic**.

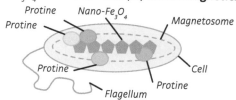

BALANCE

The **bones in our inner ears** are responsible for our **coordination**.

BIODESIGN

The field of biodesign explores ways of **designing with nature** and **working with naturally occurring processes**.

Designer **Natsai Audrey Chieza** began her research with **pigment-producing bacteria** that can **dye textiles** without producing **toxic by-products**.

FUNGI

Fungi are fantastic organisms that are neither animal nor plant. They can be multicellular or single-celled, such as yeasts.

- Evolved from protozoa about a billion years ago.
- 1.5 million species thought to exist.
- 120,000 have been taxonomically organized.
- Some can be eaten but many are poisonous.

REPRODUCTION

Also known as **zygomycetes**, fungi **reproduce sexually and asexually** by releasing **sexual spores** or **asexual sporangiospores**. They have **mating types** instead of sexes; some fungi have hundreds of mating types. Fungi **reproduction can take seconds or hundreds of years**.

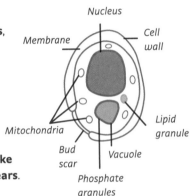

FUNGI LIFE STRATEGIES

- **Decomposers: break down matter**, including wood.
- **Mutualists**: help plants **absorb nutrients**, and **embed themselves in root tissue**, called **mycorrhizae**, vital in **ecosystems** and **agriculture**.
- **Parasites**: feed on organisms **without killing them**—until they do.
- **Predatory**: will **capture prey** with their **hyphae**.

FUNGI FEEDING

Fungi **eat decaying matter** by excreting **powerful enzymes** to **release essential compounds** into the environment. Fungi are **heterotrophs** (they cannot generate their own food).

HYPHAE

These are filament-like structures that **grow through and around whatever fungi are eating**. Hyphae contain **chitin—a polysaccharide found in shells and exoskeletons**.

MYCELIUM

A vast web of hyphae that **maximizes the surface area for absorbing food**. It is the main body of fungi and **lives underground**. As well as **nourishing the fungi**, it **maintains soil structure and health**.

FRUITING BODY

The **"sporocarp"** is the fungus's **organ for producing spores**.

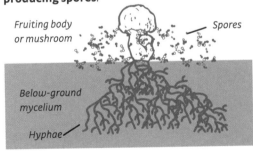

FUNGI–BACTERIAL WARFARE

Fungi and bacteria engage in **molecular warfare** as they **compete for the same resources**.

FUNGAL INFECTIONS

Fungi can infect **humans, animals,** and **plants**. Fungal infections **in crops** are a **big threat to agriculture**. Increasing **global temperatures increases the risk** of fungal infections.

THE DISCOVERY OF PENICILLIN

In the 1800s, tuberculosis caused nearly 25 percent of deaths; in the 1940s, a mere scratch from a rose could cause death from sepsis (blood poisoning).

In 1928 bacteriologist **Alexander Fleming** discovered an agar plate in his laboratory with **Staphylococcus bacteria** that had been contaminated with **mold**; he noticed that the mold created a **bacteria-free barrier**.

Scientists **Howard Florey** and **Ernst Chain** discovered Fleming's papers. During the Second World War they experimented with penicillin with biochemist **Norman Heatley**. Equipment was hard to find, so they **built apparatus** to extract penicillin **from an old bookcase and milk urns**.

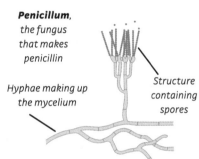

Penicillum, the fungus that makes penicillin

Hyphae making up the mycelium

Structure containing spores

In September 1940 a man called **Albert Alexander** accidentally scratched his face on the thorn of a rose; it became swollen and infected, and caused **sepsis**. Florey and Chain asked if they could test penicillin on Alexander, who **soon began to recover**. Sadly, the **penicillin ran out before Alexander could be fully cured**.

Growing and extracting penicillin was difficult, but enough was made using a **different strain of mold**. By the end of the war, **people** were **no longer dying from bacterial infections**.

DEVELOPMENT OF PENICILLIN

"Without Fleming, no Florey or Chain; without Chain, no Florey; without Florey, no Heatley; without Heatley, no penicillin."—Professor Henry Harris, 1998

ANTIBIOTIC RESISTANCE

Fleming warned that bacteria would **evolve resistance to antibiotics** if not used carefully. Antibiotics **must be used until an infection has totally cleared** and **not be overused**. Antibiotics in **farming** and **over-use by humans** has **pushed bacteria to evolve resistance**. They have:

1. Evolved enzymes that degrade antibacterial drugs;
2. Evolved changes to bacterial proteins that are used as antimicrobial targets;
3. Evolved changes in their cell membrane.

STRUCTURE

In 1945 the **structure of penicillin** was discovered by **Dorothy Hodgkin** to contain a **"lactam ring."** Penicillin **kills bacteria by binding to the cell membrane**.

Penicillin structure	R-group	Drug name
	$-CH_2-\bigcirc$	penicillin G
	$CH_2-O-\bigcirc$	penicillin V
	$-CH-\bigcirc$ CH_2	ampicillin
	$-CH-\bigcirc-OH$ CH_2	amoxicillin
	$CH_3O-\bigcirc-$ CH_3O	methicillin

ALLERGY

Penicillin can cause **serious allergic reactions** in some people.

PHOTOSYNTHESIS

Photosynthesis is the chemical process plants use to produce their food. Plants are autotrophs and can produce their own nutrients.

PLANT CELL STRUCTURE

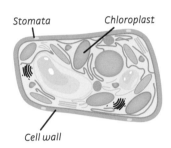

Plant cells have a **rigid cell wall** and contain **plastids**. **Chlorophyll** is the **pigment** found in the **chloroplast plastids**. Chloroplasts are the **organelles** that **absorb photons to use as energy**.

Chloroplasts have a **complex structure** containing within them a **lipid bilayer**: **thylakoids contain lumen and chlorophyll**, which are stacked into structures called **grana**; **stroba** are outside the thylakoid.

Photons that reach a plant **trigger stage one of photosynthesis** in the chloroplasts, which **breaks apart water molecules**. This leads to the **light-independent reaction**, which **completes photosynthesis**, making **sugar** and **oxygen**. The complete reaction is below:

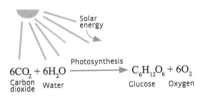

$6CO_2 + 6H_2O \xrightarrow{Photosynthesis} C_6H_{12}O_6 + 6O_2$

Carbon dioxide, Water, Glucose, Oxygen

We perceive plants as green because they reflect green light. Plants **eat only red light**. Some plants found in **deep jungles** contain **red or purple pigmentations** in their leaves; this **reflects red light back into the leaf** to **maximize edible photons**.

PLANT TISSUE STRUCTURES: C3, C4, AND CAM

Different plants have **different tissue structures** depending on what climate they grow in, and **how they "fix" (assimilate) carbon dioxide during photosynthesis** reactions.

C3: most plants; loses water through perspiration; fixes carbon in the Calvin cycle; enzyme used in the reaction: rubisco.
C4: tropical grasses; loses less water; fixes carbon in the cytoplasm; enzyme used: PEP-ase.
CAM: succulents, pineapples, cacti; water efficient; fixes carbon at night only; enzyme used: PEP-ase.

CHLOROPHYLL

Chlorophyll comes in **different forms**: **a** (contains a CH_3 molecule) and **b** (contains a CHO). Both are **stable** and **alternate single and double bonds** that **allow electron orbitals to delocalize around a central magnesium atom**. This makes them **excellent photoreceptors**.

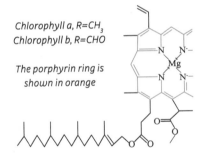

Chlorophyll a, R=CH_3
Chlorophyll b, R=CHO

The porphyrin ring is shown in orange

CYANOBACTERIA

A **diverse group of bacteria that live in water and photosynthesize to produce energy**. They are the **only autotrophic prokaryotes**. Photosynthesis evolved from **cells engulfing cyanobacteria** in a process called **"endosymbiosis."** Photosynthesis **first evolved** 2,450–2,320 million years ago.

MULTICELLULARITY

In multicellular organisms, different cells are codependent and have different specialisms that support the whole organism.

THE EVOLUTION OF MULTICELLULARITY

Single-celled organisms first appeared about 3.5 billion years ago when the Earth was a billion years old. **Multicellular organisms evolved more than once**, resulting in **different species**: **plants**, **animals**, and **fungi**. The evolution of **multicellularity** is thought to require the following to occur:
1. Cell–cell **adhesion**;
2. Cell–cell **molecular communication**, **cooperation**, and **cell specialization**;
3. **Transitions** from **"simple"** to **"complex" tissue types**.

SIMPLE ORGANISMS

DIPLOBLASTIC: Cnidaria (**jellyfish**, **corals**, **hydra**, and **sea anemones**) and other diploblastic organisms have **two germ layers**.

TRIPLOBLASTIC ACOELOMATE: Simple organisms with **three germ layers**, such as the **nematode worms**, **hookworms**, and **rotifers**.

TRIPLOBLASTIC COELOMATE: These include a tissue structure called **coelom** (a **fluid-filled cavity** that **stores and protects organ systems**)—e.g., **clams**, **snails**, and **squid**.

SEA SPONGES

These are **multicellular organisms** with **simple tissue not formed in layers** but made of a **mass of sponge cells with no specialized function**. There are around 10,000 different known species of sea sponge.

MULTICELLULAR DEVELOPMENT

- When a **sex cell** (a zygote) is **fertilized** (egg and sperm combine), it **divides and multiplies**.
- **Morphogenesis**: cells take on **specific forms**.
- **Differentiation**: cells **specialize into different types**.

COMPLEX ANIMALS

Tissue layers result in **segmentation**—e.g., **ribs**, **teeth**, **brain folds**, and **eyeballs**.

ANNELIDA: **Leeches** and **earthworms**.

ARTHROPODA: **Invertebrate insects** with **exoskeletons** and **jointed limbs**—e.g., **beetles, spiders, lobsters,** and **butterflies**.

CHORDATA: **Vertebrate organisms** with a **spinal cord**—e.g., **birds, animals,** and **fish**.

INCREASING COMPLEXITY OF STRUCTURE

- **ECTODERM**: outer layer (shell or skin).
- **ENDODERM**: digestion.
- **MESODERM**: organ tissues.
- **COELOM**: fluid-filled cavity.

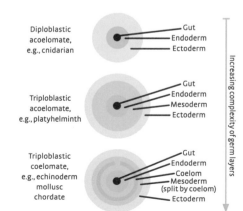

SYMBIOSIS

Symbiosis is a mutual relationship between species. Inter-species competition for resources is the driving force behind diversity in an ecosystem. One way around competition is to work together. There are symbiotic relationships everywhere.

CATEGORIES

Parasitism: benefits one species while the other is harmed.
Mutualism: benefits both.
Commensalism: benefits one organism while the other is not harmed.

PLANT INSECT POLLINATION

Flowering plants produce a **sweet nectar** that **insects feed from**. Insects move from flower to flower as they feed; as they do so they **pollinate the plant**, helping to **produce the next generation**.

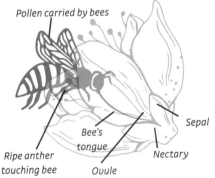

CORALS AND ZOOXANTHELLAE

Coral reefs are made from **polyps** (related to jellyfish), which have a **mouth**, **tentacles**, and a **digestive tract**. They **take in dissolved minerals**, combining them with **proteins** to slowly construct the **calcium carbonate structures** they live in. Polyp **tentacles sting and capture small organisms**, which are digested. Most **reef-building polyps** contain **zooxanthellae**—**photosynthesizing algae that produce oxygen and glucose** and feed during the day. The **glucose feeds the algae and the polyp**. The food captured by the polyps (and **carbon dioxide they release**) **feeds the algae** in return. They can feed twenty-four hours a day and, amazingly, **every new generation of corals acquire their zooxanthellae themselves**.

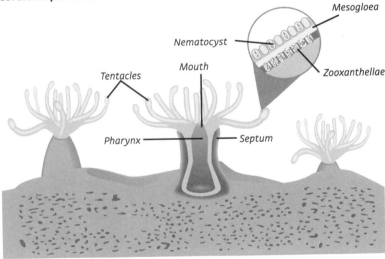

CLOWN FISH

Clown fish live in **mutual symbiosis** with **anemones** (similar to jellyfish and coral polyps). The **anemone stings protect the clown fish from predators** and **clown fish protect the anemone from smaller fish and keep the anemone clean**. Anemone **sting immunity** is gained by the **clown fish being coated by a special mucus produced by the anemone**.

MICROBIOME

Microorganisms such as bacteria, archaea, and fungi live on and in other organisms in symbiotic relationships. Without them, we could not survive. They protect us from infection, and break down the food we eat, releasing nutrients and energy.

Our bodies host billions of microorganisms, including **bacteria, fungi, single-celled organisms called archaea**, and **viruses**. The **microbiome** is the **combination of all the genes of these organisms**. Protecting the **microbiome is important to health and well-being**.

SYMBIOSIS

We have a **codependent relationship** with many microbes. They are so **crucial to health** that they are sometimes referred to as the **forgotten organ**. The **more diverse the microbiome**, the **more good health can be achieved**. **Disease** occurs **when a specific microbe takes over**—this is **infection**.

FOOD

What you eat is important to the microbiome. A diet with plenty of **fiber** helps matter move through the gut regularly. If food does not move through the intestines at the right pace, it can **decompose to the point of being toxic, killing the microbes that keep us alive**. Eating too much **fat and sugar slows digestion**, making it **difficult for microbes to thrive**.

MORE THAN HUMAN

Microbes in our body **outnumber our own cells by about ten to one.**

GUT MICROBIOME

Sometimes called the **gut flora. Without bacteria we can't digest food or access important nutrients** from the food we eat. **Different strains** of gut bacteria specialize in **different roles: creating vitamins, breaking down food**, and **protecting us from infection**.

SKIN MICROBIOME

Your **skin** is also **inhabited by bacteria, fungi, viruses**, and **archaea**. An **imbalance** of the microbiome **can lead to skin infections, autoimmune disease**, and **acne**.

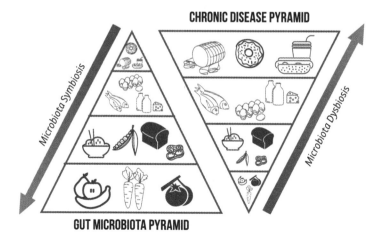

Food poisoning is the introduction of a new bacteria that **upsets the microbiome balance**.

EVOLUTION

This denotes cumulative change in the genetics and inherited characteristics of a population of organisms. Evolution is about diversity, not linear steps of ever more "complex" beings.

ON THE ORIGIN OF SPECIES

Many scientists before **Darwin** noticed that **species change over time**: from the patterns on moth wings to those of crabs. Arab scholar **Al-Jahiz** (776–868) noted **changes in competition and predation in relation to time and geography**. Darwin's contribution was the publication in 1859 of **On the Origin of Species** and the idea of **natural selection**, an idea that **Alfred Russel Wallace** (1823–1913) had also been developing.

> **NATURAL SELECTION KEY POINTS**
> - **Survival and reproduction** depend on a variety of traits.
> - **Heritable traits** in a population **change over time**.
> - **Variation** exists within all populations.

COMMON ANCESTORS

Humans and chimps share 98.4 percent of their DNA. Another misconception is to think that we are **"more evolved"** than chimpanzees; rather, we **evolved from a common ancestor** about 7 million years ago. **Whales**, **killer whales**, and **dolphins** shared a **common ancestor** with **hippos**, **cows**, and **antelopes** around 55 million years ago.

The expression "survival of the fittest" leads to misconceptions. It does not mean strongest, but rather how well an organism thrives in its environment to produce offspring.

FEATHERED DINOSAURS

Among the **survivors** of the **Triassic–Jurassic extinction** about 201.3 million years ago were **tiny feathered dinosaurs**—the **common ancestor of** the **birds** we see today. **Crocodiles** and birds share a common ancestor.

ADAPTATIONS

Species **adapt** to their environment. **Heritable traits** enable them to better suit their environment. **Organisms more adapted to survive are more likely to birth the next generation**. In 1835 **Charles Darwin** (1809–82) explored the **Galapagos Islands**. He noticed that each island had its **own species of finch** with a **beak shape adapted to the ecosystem** and food each island offered.

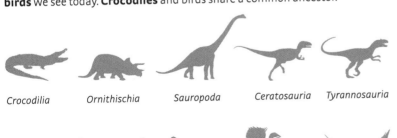

Ornithomimosauria Deinonycchosauria Archaeopteryx Birds

GENETICS AND MUTATION

If an organism dies before having any offspring, its genes are lost from the gene pool. If another organism has plenty of offspring, its genes are readily available in the gene pool.

GENE TERMINOLOGY
- **Allele**: A gene that exists in many forms.
- **Genotypes**: The gene for a specific trait.
- **Phenotype**: Physical characteristics.

ENVIRONMENTAL EFFECTS

Phenotype expression is a result of the **expressed genotype** and **environment**. In the nineteenth century, the light gray **peppered moth** camouflaged perfectly on **lichen-covered trees**. **Smog** and **pollution** in industrial areas **killed the lichen** and **turned the trees black**. Lighter grey, **less camouflaged moths became easier to prey on** and **selection pressures favored the darker moths**.

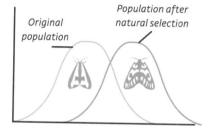

RANDOM MUTATIONS

Genetic mutations can be **useful, pointless**, and **even harmful** to organisms—mutations are **completely indifferent to their effect on an organism**. They happen at **random**, alongside **inherited adaptations** that get **passed on to each generation. Random mutations can become selected traits.**

INHERITED TRAITS

Some traits are **"recessive"**; some are **"dominant." Dominant genotypes** are **more likely to be inherited**. Phenotypic trait change—e.g., developing muscles through exercise—is **not inherited**; just because parents work out, it does not mean they will have a muscly baby. **Selection drives inheritance** within the **context of environment and heritability**.

GENETIC DRIFT

Alongside adaptations and random mutations, we get **drift**. This is a **random change in the frequency of an allele** that is **often associated with *chance* disappearances of traits in** small populations.

EPIGENETICS

Epigenetics studies **patterns of gene expression passed generation to generation**. At different *stages* of pregnancy, **stress can be inherited** due to epigenetics.

EPIGENETIC MECHANISM

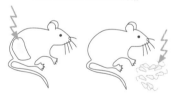

Prenatal stress   *Postnatal stress*

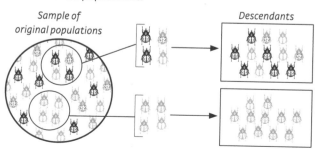

THE EVOLUTION OF PETS

Over 5,000 years, through **selective breeding**, humans turned wolves into Saint Bernards and Chihuahuas. **Inbreeding** results in **severe health problems** in all animals.

ZOOLOGY

Zoology is the study of living organisms, how they live and adapt to their habitats, compete and coexist with each other. Comparative anatomy compares the anatomies of different species.

ZOOLOGICAL CLASSIFICATIONS

Classifying living organisms is important to ecologists and scientists needing to make comparisons. **Classification schemes** reflect **evolutionary relationships.**

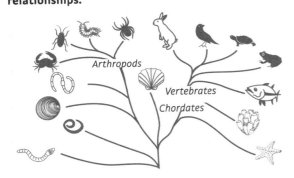

Specific species can be organized into **taxonomic categories**:

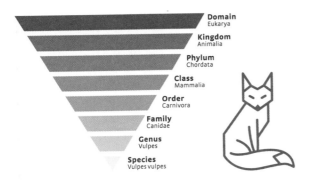

CONVERGENT EVOLUTION

Some species have **similar body parts** but are not at all the same. **Similarities** between the anatomies of **penguins** and **seals** tell us a lot about **survival in polar ecosystems and evolution**—they have a **large body size, flippers,** and **metabolic adaptations** that help them survive harsh winters.

MORPHOLOGY

This is the study of structural features such as **body size** and **shape. Dolphins** have very similar body structures to an **ichthyosaur** (extinct dinosaur). Ichthyosaurs lived in water about 250 million years ago. We can see from their fossilized remains that they had **fins**, were **predatory**, and **gave birth to live young**—just like dolphins do today. **Ichthyosaurs**, however, are **genetically more closely related to a chicken**, and **dolphins to a rabbit**.

SKELETON MORPHOLOGY

By comparing the **upper limbs** of a range of organisms, we can see how the same **bone structure** has **changed over evolutionary time to suit specific adaptations**—e.g., **swimming, flying,** and **climbing**. This supports genetic evidence of **evolutionary ancestry**.

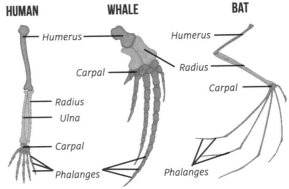

COMPARISONS AND TISSUE TYPES

Animal cells have **different body parts for moving and eating**, but once nutrients are in the body, **many animals' cells treat nutrients similarly**. Some animal cells have **special adaptations** that **affect metabolism**, such as in **hibernation** and **bird migration**.

REPRODUCTION AND CLONING

*Some species reproduce asexually. Some species engage in sexual behavior.
Twins are natural clones.*

GERM CELLS OR GAMETOCYTES

- The reproductive cells of an organism—e.g., **spores in fungi**, **egg** and **sperm in chordates**.
- Gametocytes have **one set of chromosomes** and are **"haploid."**
- **Human sex cells** contain **twenty-three chromosomes**.

SOMATIC CELLS

- Cells in an organism are **"somatic."**
- **Human somatic cells** contain **forty-six chromosomes**.
- Forty-six chromosomes is **double the "haploid" number of chromosomes**.

MATING HABITS

- Male **birds of paradise** carry out elaborate **mating dances**.
- **Snails** are **hermaphrodites** (they have both male and female organs).
- **Female angler fish fuse with males after mating.**
- **Male praying mantis** get **eaten by females**.
- **Male giraffes drink a female's urine** to tell if she is **ready for mating**.
- Some **insects** procreate **asexually**—this is **parthenogenesis**.
- Many species of fish **change sex**—e.g., **kobudai, clown fish**, and **gobies**.
- Most **female raptors** (**eagles, hawks**, and **owls**) are **larger than males**.
- **Dolphins** have **homosexual sex for fun**.

MITOSIS

Cell division whereby **cells duplicate genetic materials** and **divide in order to grow and repair tissue**.

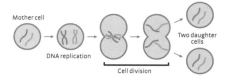

MEIOSIS

A **special type of cell division** in which **chromosomes are halved**. The sex cells of sexually reproducing **single-celled and multicellular eukaryotes**, such as **animals**, **plants**, and **fungi**, take part in meiosis.

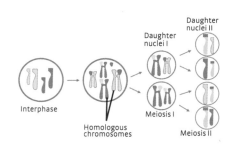

HUMANS AND SEX

From love songs, to STDs, to babies, to societal prejudice—sex is a genuine minefield. Some humans are **homosexual**, some are **heterosexual**, **bisexual**, **pansexual**, and **asexual**—all of which are **natural**. Sex for humans is **pleasurable**. Some humans are **monogamous**; some are not. The most important aspect of human sex is **consent**—both (or all) partners in a sexual relationship **MUST be old enough** and in all aspects of cognition be capable of fully consenting to participate in sexual acts.

TWINS

- **Identical twins** are **"monozygotic,"** meaning they **come from the same egg**—identical twins are **natural clones**.
- **Non-identical** twins are **born from two separate fertilized eggs**.

STEM CELLS

Stem cells have the ability to develop and specialize into any cell type in the body. They are undifferentiated cells. It is hoped that stem cell therapies already in development will one day be able to replace cells and tissues that have been damaged or lost due to injury or disease.

Most of the cells in our body—**liver cells**, **retina cells**, and **blood cells**— **have to be replaced**, and are **specialized to do a specific job**.

ADULT STEM CELLS

- **Pluripotent cells**: There are **very few of these in the body**; they reside in different tissues— e.g., **teeth**, **bone marrow**, **blood vessels**. A pluripotent cell will **differentiate into whatever cells are around it**.
- **Multipotent cells**: These are **more common**, but **more restricted in what cells they can become**.

EMBRYONIC STEM CELLS

These are the **cells in an embryo in its very early stages of development**. When an egg is fertilized by sperm, the **diploid number of chromosomes is complete** and the **cell undergoes meiosis** (divides and multiplies). It gets to a point where it is called a **blastocyst**. Within the blastocyst are the embryonic stem cells.

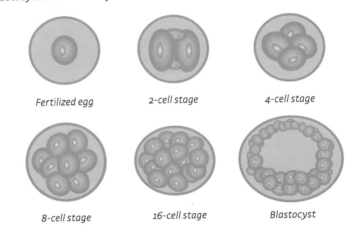

Fertilized egg — 2-cell stage — 4-cell stage — 8-cell stage — 16-cell stage — Blastocyst

IN VITRO FERTILIZATION

In vitro fertilization (**IVF**) is **used to help some people to conceive**. It is **also used** (with donor permission) **to carry out stem cell research**.

STEM CELL RESEARCH

While **few stem cell therapies** have proved both **safe and reliable, advances are being made** regarding the use of stem cells **to repair and replace diseased or damaged cells**. **In principle, stem cells can** be brought to a liver to **replace liver cells**, but keeping the stem cell in place is problematic.

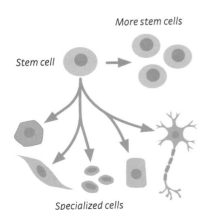

Stem cell — More stem cells — Specialized cells

SYSTEMS OF THE BODY

Human anatomy is divided into eleven main systems, which all interact with one another.

SYSTEM SPECIALISMS

- **Cardiovascular**: the **heart**; **veins** and **arteries**; **red blood cells** that carry oxygen and have no nucleus; **white blood cells**; **plasma** and **platelets**.
- **Respiratory**: **alveoli** in the **lungs** allow CO_2 and O_2 to permeate during **gas exchange**.
- **Digestive**: includes the whole digestive tract from **mouth** to **anus** and the **liver**.
- **Renal**: includes the **kidneys** and **bladder**, which filter **toxins** from the **blood**.
- **Nervous**: cells can send and sense **electrical signals**.
- **Endocrine**: includes the system of **hormones** our body needs to **self-regulate**; includes the **adrenal** and **pituitary glands**, **pancreas**, **ovaries**, **thyroid**, **brain**, **testicles**, and **thymus**.
- **Immune**: white blood cells, i.e., lymphocytes (T-cells, B-cells, and NK cells); neutrophils; monocytes/macrophages; and includes **spleen** cells.
- **Integumentary**: **hair**, **nails**, and **skin cells**; layers include **oil glands**, **fat cells**, **skin cells** with **melanin**, **sweat glands**.
- **Skeletal**: **osteogenic** (bone-making) cells develop into **osteoblasts**, which create the **biomineral bone matrix** and form **osteocytes**.
- **Muscle**: these cells have **"contractile" proteins** that help muscle stretch.
- **Reproductive**: **genitals** and **inner reproductive organs**.

COMPLEXITY OF BONE TISSUE

COMPLEXITY OF SKIN

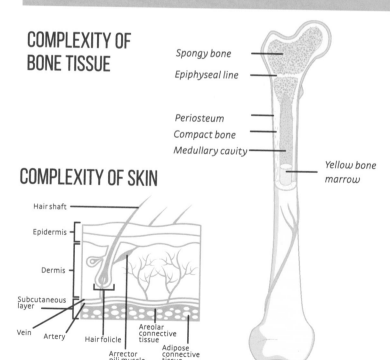

CELL SPECIALIZATION

There are over 200 different types of cells in an adult human. Organisms made of many different cells begin life as a **zygote** (egg fertilized by sperm). It takes four days for a human zygote to become a **blastocyst**, and a previously homogenous ball of cells divides and multiplies, beginning to **specialize** into specific cells, i.e., organ tissue cells.

GENE EXPRESSION

Each **specialized somatic cell** contains the same **DNA**, yet each cell **needs to read the correct section of DNA** in order to make it **synthesize the correct proteins** for it **to do its job**. When a cell is actively using specific genes, it is said to be **expressing these genes**.

HUMAN ANATOMY

Anatomy is the study of the body in terms of organ systems, areas of the body, different tissues, and how they interrelate.

APPROACHES TO ANATOMY

- **Systemic approach**: the study of systems.
- **Regional approach**: the study of areas.

ANATOMICAL PLANES

Anatomists use **body planes** to discuss **specific structures** within areas of the body.

SURFACE ANATOMY

Primarily concerned with the skin and **musculoskeletal system**. Surface anatomy is concerned with what anatomy can be understood from examining the body externally.

NAMING MUSCLES

The names of muscles sound very complicated, but some basic rules are applied:

- Size
- Shape
- Location
- Orientation of muscle fibers
- Actions of muscles
- Points of origin of a muscle
- Points of origin and insertion
- Muscle function

MOVEMENT

Anatomists use specific words to describe how muscles move in relation to each other.
Protraction: anterior movement
Retraction: posterior movement
Abduction: moves away from
Adduction: moves toward
Flexion: movement toward
Extension: movement away
Pronators: turn down or back
Supinators: turn up or forward
Levators: lift a limb
Depressors: push a limb down
Rotators: rotate
Sphincters: open or close in a ring shape

SIZE OF MUSCLES

Maximus or magnus: the largest muscle in the group
Minimum: the smallest
Longus:: the longest
Brevis: the shortest
Latissimus: the widest

SHAPE OF MUSCLES

Trapezius: trapezium shaped
Deltoid: triangular
Serratus: "serrated"
Platysma: flat and broad

ORIENTATION OF MUSCLES

Rectus: fibers run parallel to midline or spine
Oblique: at an angle
Transverse: across

NUMBER OF ORIGIN POINTS FROM BONE AND INTERSECTION

Bi: two
Tri: three
Quad: four
(Example: **biceps, triceps**, and **quadriceps**.)

FUNCTION

Examples of function include:
Masseter: chewing muscles
Risorius: smiling muscles

LOCATION

Medial: near the middle
Lateral: outer

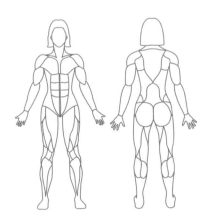

IMMUNOLOGY

The immune system is the body's defense against infectious organisms.

ANTIGEN/PATHOGEN

An **organism** or **particle** that **stimulates an immune response**.

ANTIBODIES

Antibodies are **proteins** that **stick to pathogens**. **Immunoglobulin** is a large, Y-shaped protein **produced by plasma cells** that **neutralizes pathogens**.

LYMPHOCYTES

Lymphocyte cells are **white blood cells**. They **kill and eat pathogens** in a process called **phagocytosis**. They include: **T-cells, B-cells, NK cells, neutrophils, monocytes**, and **macrophages**.

ADAPTIVE IMMUNITY

The **immune system** is **activated** when a **virus** or **bacteria** manages to enter the body. Cells called **B-cells** and **T-cells**, specialized in attacking **antigens**, are able to recognize **microbes** they have **encountered before**.

ANTIBODY

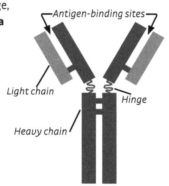

Antigen-binding sites
Light chain
Hinge
Heavy chain

ANTIGEN

- A protein on a foreign object that stimulates the immune system to produce antibodies
- A virus, bacteria, toxin, etc., that triggers the immune response

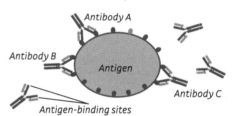

Antibody A
Antibody B
Antigen
Antibody C
Antigen-binding sites

IMMUNE SYSTEM SITES

The immune system is situated in different parts of the body:
- **Lymph nodes**: lymphocyte cells live in the lymph nodes and can recognize pathogens.
- **White blood cells**: attack pathogens.
- **Bone marrow**: this is where blood cells are produced.
- **Villi in the lungs**: physically remove or obstruct pathogens.
- **Skin**: forms a barrier.
- **Stomach**: stomach acid kills bacteria.
- **Spleen**: protects against bacterial infection.

IMMUNE STUDIES

- Cancer immunotherapy
- Immune regulation
- Viral immunobiology
- Inflammation research
- Tumor immunology

IMMUNE SYSTEM DISEASE

- **Autoimmune: overactive immune system**, the **body attacks and damages its own tissues**. (Examples: **allergic reactions, arthritis, type 1 diabetes, psoriasis, celiac disease, lupus, narcolepsy**.)
- **Immune deficiency: decreased ability to defend the body from pathogens**, resulting in **vulnerability to infections**.
- **Cancer:** cancers use clever molecules to **hide from the immune system**; cancer is **uncontrolled cell growth** in one's own body.
- **Hygiene hypothesis:** being **too clean** interrupts immune function.

CIRCULATION OF THE BLOOD

Blood is a complex liquid. Half of it is plasma (made of water salts and protein). It contains red blood cells, white blood cells, and platelets.

BLOOD CELLS

Most red blood cells, white blood cells, and platelets are made in the **bone marrow**.

- **White blood cells (leukocytes)**: the main different kinds are **B-cells** and **T-cells**; white blood cells are part of the **immune system**.
- **Platelets**: help the blood to **coagulate** (thicken) when the skin is cut or punctured, causing a **scab** to form.
- **Red blood cells (erythrocytes)**: transport **oxygen** around the body; have a **structure** that **helps to maximize the surface area they can use to absorb oxygen**; contain a protein called **hemoglobin** that binds to oxygen; they **cannot undergo meiosis** as they have **no nucleus**.

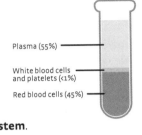

CIRCULATION OF THE BLOOD

English physician **William Harvey** (1578–1657) discovered that the **heart pumps blood around the body**.

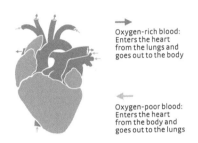

Oxygen-rich blood: Enters the heart from the lungs and goes out to the body

Oxygen-poor blood: Enters the heart from the body and goes out to the lungs

He noticed that **veins** have special **v-shaped valves** that make the **blood circulate in one direction** and recognized that **arteries and veins have different functions**:
- **Arteries**: carry blood to the rest of the body, **away from the heart**.
- **Veins**: carry blood from organs, **toward the heart**.

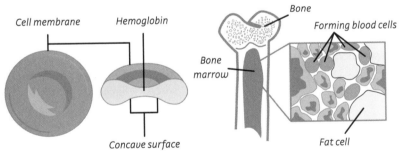

DOUBLE SYSTEM

The human circulatory system is a **double system**: one **between the heart and lungs**; one **between the heart and other organs**.
- **Pulmonary circuit**: carries blood **to the lungs**, where it is **oxygenated** through **respiration**.
- **Systemic circuit**: blood is pumped **around the body**, bringing **oxygen** and **nutrients** to be **diffused into tissues**.

RESPIRATION

- **Breath in**: **oxygen** diffuses in the **alveoli** in the **lungs**; **hemoglobin** in the blood **absorbs oxygen**.
- **Breath out**: **carbon dioxide** diffuses from the **alveoli**, **exiting the blood**.

PARASITOLOGY

Parasites can be single-celled or multicellular organisms. Parasitology is the study of parasites.

ECTOPARASITES AND ENDOPARASITES

Ectoparasites infect their host **on the skin or outside**; **endoparasites** on the **inside**.

OPHIOCORDYCEPS

These are **parasitic fungi** that **infect insects**, feeding within them peacefully until they need to **release spores**. They then **control an insect's brain**, moving the creature to a moist location, whereupon they **burst out of its head**.

PARASITOID WASPS

- These **lay their eggs on or in the bodies of other arthropods**; e.g., **caterpillars**.
- **Eggs** develop and **hatch into larvae** that **control the development** of the caterpillar by **releasing hormones**.
- When the larvae are ready, they **paralyze the caterpillar**.
- Endoparasitic wasp larvae **burrow out of the caterpillar's body**.

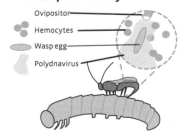

WOLBACHIA

A **bacterial parasite** that infects **armadillidium (pill bugs, ladybugs)** and spreads via the gametes—the egg cell in this case. Wolbachia **turns genetic males into females** so as to spread into more eggs. The sex ratio of the insects becomes distorted, but **armadillidium can become intersex** to avoid extinction.

PARASITE FOR GOOD

Wolbachia also affects **mosquitoes** and is being used to help manage the **Zika virus, dengue fever,** and **yellow fever** by **competing against viruses for nutritious molecules** in the mosquito, making it harder for viruses to grow.

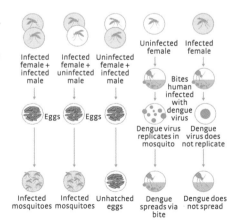

PLASMODIUM

- Responsible for **malaria**.
- Cycles from infecting mosquitoes to **humans**.
- Infects the **liver** and **red blood cells**, causing **fever, tiredness, vomiting, headaches, seizures**, and **death**.
- In 2017 alone, 435,000 deaths resulted from malaria.

TU YOUYOU AND THE DISCOVERY OF ARTEMISININ

Tu Youyou is a Chinese pharmaceutical chemist born in 1930. Following extensive research, Tu Youyou used **techniques from traditional Chinese medicine** to extract **artemisinin** from **sweet wormwood**—a compound that **inhibits the plasmodium**. She was awarded the **2015 Nobel Prize in Medicine** for her work.

THE ARTEMISININ MOLECULE

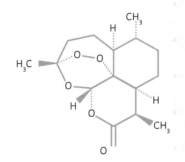

NEUROSCIENCE

Neuroscience explores the nervous system, the emergent properties of neurons, and the connections between them, cognition, and consciousness.

- **Central nervous system** (**CNS**): **brain** and **spinal** cord.
- **Peripheral nervous system** (**PNS**): **nerves**—**sensory**, **motor** (somatic and autonomic).
- **Enteric nervous system**: the **nervous system of the gut** (independent of the CNS).

OPERATION

- **Sensory input**: detects **changes in environment**—**impulses transmitted** from sensory nerves toward the **CNS**.
- **Integration**: **processes the changes** and **decides what to do**.
- **Motor output**: **response** to the input—impulses **toward muscle**.

NEURONS

- **Nerve cells** have a specialized structure that **receives and sends electrical signals**.
- They are the **longest-living** cells in the body.
- Most are **amitotic** (meaning they do not replace themselves).
- **Axons** transports **signals**.
- Neurons **connect and form neural networks**.

Diagram labels: Dendrite, Node of Ranvier, Cell body, Axon terminal, Schwann cell, Myelin sheath, Axon, Nucleus

GLIAL CELLS

These support neuron function:

CNS
- **Astrocytes**: regulation, support.
- **Microglial cells**: protect spinal cord.
- **Ependymal cells**: brain lining.
- **Oligodendrocytes**: create myelin sheath.

PNS
- **Schwann cells**: insulate the myelin sheath.
- **Satellite cells**: surround nerve cells.

ACTION POTENTIAL

The *frequency* of the **electrical signal** in a neuron communicates the *intensity* of the message; e.g., **big pain = high frequency**.

BRAIN

Specific areas of the brain are responsible for different functions:

- **Cerebrum**: made of folds of white matter enveloped in grey matter.
- **Frontal lobe**: emotions, priorities, planning, problem solving.
- **Motor cortex**: movement.
- **Temporal lobe**: memory, languages.
- **Sensory cortex**: senses.
- **Parietal lobe**: perception, understanding, logical processing.
- **Occipital lobe**: visual, spatial.
- **Cerebellum**: coordination.
- **Corpus callosum**: connects both sides of brain.
- **Brainstem**: midbrain, pons, and medulla oblongata—relays information; regulates heart, breathing, sleep, pain, awareness of stimuli.
- **Diencephalon**: hypothalamus, epithalamus, mammillary body, and limbic system—reproduction, safety, eating, drinking, sleep, strong emotions such as fear.

SURGERY

Some diseases and/or injuries can be treated only by surgical means: by creating an incision through the skin and moving, removing, repairing, or altering parts of the body from the inside and by using specialized instruments.

ANESTHETIC

Before anesthetic, the **pain alone from surgery would kill patients**—particularly with **tooth removal**—a procedure that more people needed.

ANTISEPTIC

This helped to **lower infection rates**. In cases of **deep surgery**, antiseptic is **dangerous and cannot be used**.

ANTIBIOTICS

Before antibiotics, surgery held a **high risk of death from infection**. Thanks to antibiotics, infection is manageable, making **surgery safer**.

IMMUNE RESPONSE

Organ transplants are risky, as the body's **immune system often rejects unfamiliar tissue**. Improvements in **matching donors and patients**, and **immunosuppressant drugs**, have helped to **improve the success rate** of organ transplants.

SURGICAL TERMINOLOGY

-tomy: something is cut
-ectomy: is cut out
-ostomy: an opening is made
-plasty: is reshaped
-plexy: is moved to the correct position
-rraphy: is sewn up, e.g., gastrorrhaphy
-desis: two things are connected

Surgical instruments are organized in term of what they need to do:

- **Graspers**: grasp things.
- **Clamps and occluders**: clamp things closed; e.g., a vessel.
- **Retractors**: spread tissue.
- **Mechanical cutters.**
- **Dilators and speculae**: open an opening.
- **Suction tubes**: drain fluids.
- **Irrigation and injection needles**: inject or remove fluid.
- **Scopes and probes**: for seeing/measuring changes.

IMPLANTS

These include: **hip and knee replacements, dental implants, cosmetic and reconstructive implants**, and **electronic implants** such as **pacemakers, blood-sugar-level sensors**, and **neurostimulators**.

Syringe   *Gauze*   *Tweezers*   *Forceps*

INNOVATIONS TIMELINE

1914 First non-direct blood transfusion
1950 First kidney transplant
1960 First hip replacement
1963 First liver transplant
1964 Laser eye surgery introduced
1967 First heart transplant
1987 First heart and lung transplant
2005 First face transplant surgery
2008 Keyhole surgery using lasers performed
2011 First leg transplant
2012 First womb transplant

LIFE HISTORY

The study of "life history" evaluates the different methods (or strategies) used by different species in response to variations within their environment.

Life history and strategies events include: **life span**, **number of offspring**, **egg size**, **parental behavior**, **age of maturation**, and **age at death** in relation to the resources available in the environment.

COMPARISON OF SIZES OF DIFFERENT SPECIES

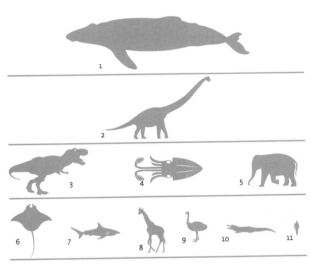

1. Blue whale
2. Diplodocus
3. Tyrannosaurus rex
4. Giant squid
5. Elephant
6. Manta ray
7. Great white shark
8. Giraffe
9. Moa
10. Crocodile
11. Human

FISH PARENTING

Two examples of parenting strategies of different fish:
- **Jawfish** protect their offspring after they have hatched from their eggs.
- **Atlantic salmon** will bury fertilized eggs in loose gravel, where they develop over the winter months. When salmon eggs hatch, they develop independently of their parents.

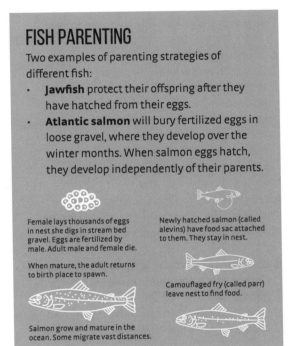

Female lays thousands of eggs in nest she digs in stream bed gravel. Eggs are fertilized by male. Adult male and female die.

When mature, the adult returns to birth place to spawn.

Salmon grow and mature in the ocean. Some migrate vast distances.

Newly hatched salmon (called alevins) have food sac attached to them. They stay in nest.

Camouflaged fry (called parr) leave nest to find food.

Life history is about the **structure of populations of organisms**. It examines the events that happen in an organism's life that ensure its survival and that of the next generation.

Population ecology is the **study of population dynamics** over time and in relation to organism populations and their environment.

Organisms can **optimize their productivity through behavior**—such that the next generation is nurtured to maturation. This is important to life history studies.

- When the environment is **harsh**, organisms need to **adapt** quickly—a **fast life history strategy** is optimal because it is more likely that they will die before their first reproductive event.
- When the environment is **good** and provides a species with all the resources necessary to survive and raise offspring safely and successfully, a **slow life history strategy** is preferable.

REPRODUCTIVE STRATEGIES
- **Semelparity**: one reproductive event before death.
- **Iteroparity**: multiple reproductive events before death.
- Some animals will **vary between semelparity and iteroparity**, depending on environmental constraints and resources.

PRINCIPLES OF ECOLOGY

Ecology is the study of different organisms in relation to one another and their environment.

1. Individual organisms make up populations. The **abundance** (number of individuals in a population) and **diversity** (number of different kinds) can **vary over time** and in relation to each other.

2. All **energy** (food) ultimately comes from the **Sun**. Carbon-rich sugars such as **carbohydrates** and **glucose** are made in **plants** and **algae** through **photosynthesis**. The movement of nutrients is best seen as a **web of interactions** rather than a direct food chain.

3. Chemical reactions in organisms are what provide them with their energy. **Chemistry and physics define how metabolism works.**

4. Chemical **nutrients cycle through ecosystems**. Elements such as **carbon, nitrogen, phosphorus,** and **sodium** cycle back and forth; this process includes the **recycling and decay of dead organisms** in an ecosystem.

5. The rate of increase in a population is controlled by the number of **births, deaths, and net migration** to an area.

6. The **diversity of an area** is determined by the number of new forms arising, those migrating into an area and those going extinct.

7. Different organisms affect one another in many ways: by **eating one another, being one another's food, sharing the same space or area, or eating the same foods. Organisms' interaction** within a geographical area affects their abundance.

8. Ecosystems are organized into **webs of interaction**, making them very complex.

9. **Human populations have a disproportionately large role in ecosystem interactions**, disrupting and shifting webs of interaction and nutrient cycles that have generally been stable for millions of years.

10. Natural processes are **essential to human life**. Ecosystems also provide materials, processes and substances (called **ecosystem services**) that humans depend on.

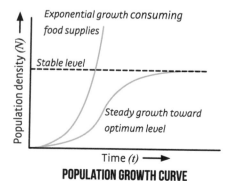

POPULATION GROWTH CURVE

WHAT DO WE GET FROM ECOSYSTEMS?

PROVISIONING SERVICES	Food	Wood	Medicine
REGULATING SERVICES	Water filtration	Crop pollination	Disease control
CULTURAL SERVICES	Spiritual	Personal growth	Leisure and fun

TROPHIC CASCADES

From predators to plants to bacterial spores, even slight changes in an ecosystem can accumulate and cause significant changes to ecosystems. This is known as a trophic cascade.

Trophism: a biological phenomenon where **change or growth occurs in response to an environmental stimulus** such as roots growing down and leaves growing toward the Sun.
Trophic level: a level in a **food chain** or area in a **food web**.

TROPHIC CASCADE

When an organism in a food chain or food web either **dies out** or is **overly populous**, it **can affect entire ecosystems**. The introduction or removal of a **predator** can have a **cascading effect** on ecosystems, altering **populations** in relation to one another and shifting **nutrient cycles**.

Homeotroph: organisms that generate their own food sources, e.g., plants directly using sunlight to create glucose.
Heterotroph: organisms that can get the nutrients they need only by eating other organisms.

WOLVES IN YELLOWSTONE

In the 1900s, people **hunted gray wolves for sport**, intensely, and without thought to what impact this might have on the ecosystem. In 1926 the last two wolves were killed, making them locally extinct. **With the wolves not around, the deer population grew.** The expanding population of deer consumed more and more shrubs, grasses, and plants. Plant life is essential for insect populations and for maintaining the quality of the soil. The **soil quality changed as a result, affecting the trees and all members in the ecosystem**. Yellowstone suffered throughout the twentieth century because of this. Wolves were **reintroduced** to Yellowstone National Park in 1995 in an attempt to restore the environment. Following wolf reintroduction, Yellowstone has **recovered, balancing populations, plant life, and soil quality**.

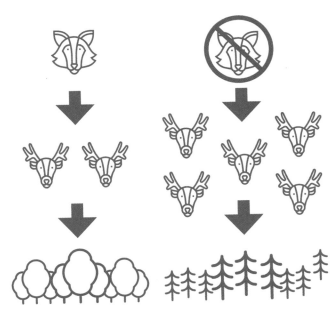

 Top-down effects: trophic levels controlled by predator population, e.g., carnivores keeping herbivore numbers in check.

 Bottom-up effects: resources such as grasses and plants restricted, impacting on herbivore numbers and, by default, carnivore numbers.

THE EARTH'S OCEANS

Seventy-one percent of the Earth is covered in water, 50 to 80 percent of all life on Earth lives in the oceans, and only 10 percent of the oceans have been explored.

OCEAN CURRENTS

Ocean currents form **definite intricate paths**. Non-uniform heating causes **convection currents**: **circulating water and nutrients**.

- Ocean currents can be thousands of miles in length.
- Cooler, denser **polar currents** sink and flow toward the warm **equator**.
- Cool water warmed by the equator then rises.
- Warmer, less dense **equatorial currents** rise and flow toward the cool **poles**.
- Warm water cooled by the poles then sinks.

CORIOLIS EFFECT

This is an effect caused by the **direction of the rotation of the Earth**.

- Winds in the northern hemisphere force currents clockwise.
- Winds in the southern hemisphere force currents counterclockwise.

SALINITY

- Salinity is the amount of **dissolved salt** in oceans.
- Salt water **is more dense than freshwater**.
- Seas and **fresh rivers** affect ocean currents.
- **Evaporation** increases salinity.
- Salinity currents carry waters of higher salinity to the bottom of the ocean (or sea) creating **deep-water currents**.

Thermohaline (heat and salt) circulation causes deep ocean currents driven by **density gradients**.

LA NIÑA / EL NIÑO

Complex oscillating changes in the direction of currents in the **central and eastern tropical Pacific Ocean**.

- **El Niño: warmer than average** sea surface temperatures, reversing wind patterns across the Pacific.
- **La Niña**: irregular intervals of **cooler-than-average sea** surface temperatures in equatorial Pacific regions.

OCEAN DEPTH

Oceanographers and **marine biologists** distinguish between different depths in the oceans.

THE MARIANA TRENCH

This is a vast trench at the **very bottom of the western Pacific Ocean**. It is 1,580 miles long and 43 miles wide. Its deepest point is 36,201 ft. and **could engulf Mount Everest**.

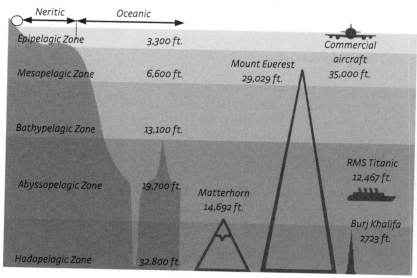

EXTINCTIONS

Extinction occurs when an organism or group of organisms completely dies out. The time it takes for extinctions to occur varies.

ENDANGERED SPECIES

Organisms threatened with extinction. The last male **northern white rhino** died in 2018.

CRITICALLY ENDANGERED SPECIES AND ESTIMATED POPULATION SIZE

Amur leopard: 60
Cross River gorilla: 250
Black-footed ferret: 300
Amur tiger: 450
Black rhino: about 5,000
Asian elephant: 40,000–50,000
Orangutan: 104,700

ENDANGERED SPECIES

Hawksbill turtle
African wild dog
Galápagos penguin
Whale shark
Chimpanzee
North Atlantic right whale

MEGAFAUNA EXTINCTION

Megafauna are big animals vulnerable to **habitat destruction**. Many of the world's megafauna are endangered.

HABITAT DESTRUCTION

Habitat destruction is one of the main causes of extinctions and **reduces biodiversity**. Causes include: **deforestation**, **pollution**, **hunting**, and **extreme natural disasters**.

INSECT POPULATION

Forty percent of the world's insects are threatened with extinction. Insects provide **essential ecosystem services**, e.g., **purifying water** and **pollinating crops**. **Insecticides** and **habitat destruction** are causing this extinction.

THE FOSSIL RECORD

The fossil record contains the preserved remains of species that are now extinct.

MASS EXTINCTIONS

Over 90 percent of all organisms that have ever lived on Earth are now extinct. **Mass extinctions can be identified geologically.**

	Extinction event	Time
1	Holocene extinction	Present
2	Cretaceous–Paleogene extinction	65 million years ago
3	Triassic–Jurassic extinction	199 million to 214 million years ago
4	Permian–Triassic extinction	251 million years ago
5	Late Devonian extinction	364 million years ago
6	Ordovician–Silurian extinction	439 million years ago

ANTHROPOCENE

Human activity is extensively altering the Earth's geophysical strata, so much so that many experts call our era the Anthropocene. The world is also experiencing its sixth mass extinction and fourth industrial revolution.

DIVERSITY AND POPULATION

Diverse ecosystems are more stable and healthy, as they offer a greater number of genes to draw on in response to changes in the environment. Diverse ecosystems and communities are more able to adapt to change.

- **Genetic diversity**: the number of genes in the DNA of a species.
- **Genetic variability**: the tendency of genes to vary.
- **Ecosystem diversity**: a type of biodiversity.

PHENOTYPE

The observable/measurable characteristics of an organism. **Eye color** is a phenotype, and so is an **ant's understanding of how to build a nest** and the **markings on the wings of a moth**. Phenotypes include:

- Physical form (morphology);
- How it develops over time;
- The biochemistry of an organism;
- Behavior and instincts.

MONOCULTURES

Diverse populations are **more resistant to disease and predation** due to having **more genes**. Take, for example, a **monoculture (genetic clones)** of a crop being grown on a large farm: a **pathogen** could easily evolve to attack that monoculture. The **lack of genetic diversity** makes the crop **more vulnerable to disease**, as it is easy to spread. Breeding monocultures of **disease-resistant crops** is a **short-term solution**, since **pathogens are constantly evolving**. In a diverse population it is harder for pathogens to spread.

MVP

Minimum viable population (MVP) is the **smallest number of individuals necessary for a species to survive in the wild**.

DIVERSITY OF FINCHES IN THE GALAPAGOS

GENOTYPE

Genotype is the name given to genes responsible for a given trait.

- **Allele** is a possible form of a gene.
- A diploid organism is **heterozygous**.
- **Wild type** is a gene as it would occur in nature.

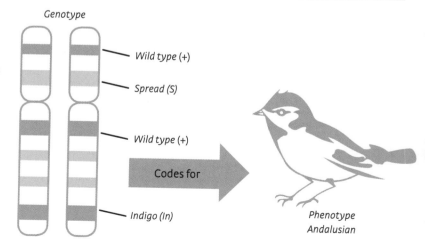

PLATE TECTONICS

Planet Earth is made of defined layers, with a comparatively thin, solid, rocky outer crust, made of tectonic plates floating on currents of churning hot rock that cause geophysical activity.

THE CRUST

- The Earth's crust is between three and twenty-five miles **thick**.
- The crust is **solid**.
- Contains **oxygen**, 46.6 percent by weight; **silicon**, 27.7 percent; **aluminum**, 8.1 percent; **iron**, 5 percent; **calcium**, 3.6 percent; **sodium**, 2.8 percent; **potassium**, 2.6 percent; and **magnesium**, 2.1 percent.

MANTLE

- 1,800 miles thick.
- Is **liquid**.
- Made up of different layers; contains **silicates**, **calcium**, **magnesium**, **iron**, and other **minerals**.
- Temperature range is 392 °F–7,230 °F.

OUTER CORE

- 1,370 miles thick.
- Is **liquid**.
- Contains mainly **iron** and **nickel**.
- Temperature range is 7,952 °F–11,012 °F.

INNER CORE

- 760 miles thick.
- Is **solid**.
- Contains mainly **iron**, **nickel**, and some **uranium**.
- Temperature up to 10,800 °F.

HEAT FLOW IN THE EARTH

Tectonic plates move about **one to two inches per year** due to underlying **magma currents**.

Heat transport mechanism

 Advection

 Convection

 Conduction

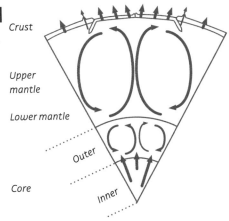

TECTONIC MOVEMENTS

Slabs of solid crust **can push against each other to form mountain ranges**. However, plates can also be **pushed beneath each other, forming a trench** in a process called subduction.

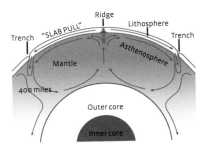

SEAFLOOR SPREADING

The **Mid-Atlantic Ridge** is a mountain range that runs for 10,000 miles along the ocean floor. At the center of the ridge, new ocean floor is forming, pushing the plates away from the ridge.

TECTONIC BOUNDARIES

- **Convergent**: plates move into one another; sometimes resulting in **subduction** or **mountain formation**.
- **Divergent**: plates move apart, forming a **ridge**.
- **Transform**: plates slide against one another.
- **Earthquakes** and **volcanoes** occur at tectonic boundaries.

ATMOSPHERIC PHYSICS

The atmosphere shields the Earth from cosmic rays, regulates the temperature on Earth, and enables us to breathe. The atmosphere contains mainly oxygen and nitrogen.

Constituent	Percent by Volume
Nitrogen (N_2)	78.084
Oxygen (O_2)	20.946
Argon (Ar)	0.934
Carbon dioxide (CO_2)	0.036
Neon (Ne)	0.00182
Helium (He)	0.000524
Methane (CH_4)	0.00015
Krypton (Kr)	0.000114
Hydrogen (H_2)	0.00005

The **atmosphere** has six main **layers**:
- **Troposphere**: **weather** and **human activity** happens in this layer.
- **Ozone**: a thin layer of **ozone**, a form of molecular oxygen, O_3.
- **Stratosphere**: no weather occurs here or past this layer.
- **Mesosphere**: this is where **asteroids** and **meteors** are typically seen burning up.
- **Thermosphere**: **aurora** happen in this layer.
- **Exosphere**: where **satellites** orbit.

UNEQUAL HEATING

The Earth is **tilted on its axis**, which is why we have **seasons**. This results in unequal heating of the atmosphere. At the **polar extremities** of the Earth there is more **diffuse heating**, while **equatorial regions** are **heated intensely**.

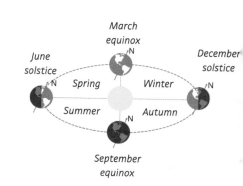

CORIOLIS EFFECT

The Earth **rotates counterclockwise** as viewed from the North Pole. This motion, along with **convection**, causes **atmospheric circulation** called **cells**. Much like ocean currents, **winds** move in **reliable patterns of movement**, such as northeasterly and southeasterly **trade winds**.

ALBEDO

This denotes **reflectivity of the Sun's rays on the Earth's surface**. The more reflective a surface is, the more **heat** is reflected back into the atmosphere.

HADLEY CELLS

The weather forms recognizable **convection currents** called **cells**:
- Cells at the **equator** create large convection currents; hot air quickly flows upward but quickly cools and condenses to form **clouds** and **precipitation** (**rain**).
- As we move toward the **poles**, the cells get smaller and smaller.

PRESSURE

- When circulating cells cycle toward each other, the result is a **decrease in pressure**.
- When circulating cells cycle away from each other, the result is an **increase in pressure**.

BIOGEOCHEMICAL CYCLES

The movement of nutrients through an ecosystem is essential to life and biodiversity. Nutrient cycles are repeated pathways taken by specific nutrients.

BIOGEOCHEMICAL CYCLES

These take place in the biosphere and incorporate different areas: **lithosphere** and **geosphere** (land) → **hydrosphere** (water) → **atmosphere** (air).

Nutrient cycles require the **circulation of elements** through **biogeological processes**.
- **Biomass**: the mass and number of living organisms; lives, excretes, and dies.
- **Litter**: dead leaves and the decaying remains of organisms.
- **Soil**: materials in the upper layer of the Earth; plants, insects, animals, and mycelium grow in soil; soil contains organisms, rock particles, decaying matter, water, and minerals.
- **Assimilation**: absorption and digestion of food or nutrients.
- **Dissilation**: decomposition of complex compounds into simple substances.

OXYGEN

- O_2 is used in **respiration** by organisms.
- O_2 is assimilated in **organic compounds**, e.g., in **proteins**, **fats**, and **sugars**.
- O_2 is excreted by plants during **photosynthesis**.
- The **hydrological cycle** is very closely linked to the **oxygen cycle**.
- **Oxidation** and **reduction reactions** play a role in the **movement of oxygen**.

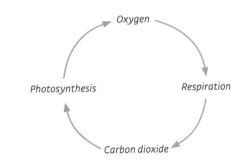

NITROGEN

Nitrogen is used in **protein synthesis**. **Bacteria** are essential to this cycle.

1. Nitrogen-fixing bacteria **convert nitrogen gas (N_2) to ammonia (NH_3)**.
2. Ammonia is converted into nitrite ions in the soil in **nitrification**.
3. Living organisms excrete ammonia in **urea**, **sweat**, and **waste**; **ammonification** takes place when bacteria **convert nitrogen-rich waste into simpler molecules**.
4. Denitrifying bacteria **convert simple nitrogen into nitrogen gas (N_2)** to be cycled again.

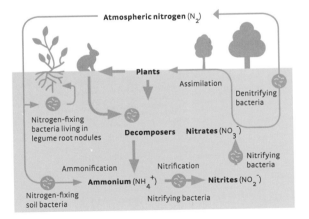

All cycles involve the **assimilation by biomass** and **dissimilation through decay** that feeds back into the cycle:
- **Phosphorus cycle**: Phosphorus is a vital element needed in **cell metabolism**.
- **Sulfur cycle**: Sulfur is important in the **formation of proteins and enzymes**.

HYDROLOGICAL CYCLE

The hydrological cycle is ultimately powered by sun's solar energy, which continually circulates the water on Earth.

Sun heats water in **oceans**, **lakes**, **rivers**, and across the Earth's surface. Water then **evaporates**; as it rises, it **cools** and **condenses** into **clouds** and forms **rain**.

Transpiration: Evaporation of water.
Condensation: Warm air rises and cools as it does so, becoming concentrated with moisture.
Precipitation: Warm air that cools forms clouds, which eventually precipitate into rain or snow. This is the main input in the water cycle. Intensity, duration, and frequency of rain or snow affects the cycle.
Surface runoff: Water on land eventually makes its way to the ocean via rivers or seeps through soils. Water can be stored in lakes, basins, and underground channels. It is sometimes stored in soil and rock. Runoff is the total amount of water flowing into a basin or reservoir.
Groundwater flow: Water is transferred through soil and rocks toward rivers and oceans.
Infiltration: Water seeps downward through the soil to reach the ocean.
Percolation: Water can also penetrate permeable rocks and percolates as groundwater.

CLOUDS

Clouds form **different shapes at different altitudes**. **Air pressure** and **temperature** decrease as the altitude increases. All weather takes place within the **troposphere**. Clouds are part of the **water cycle** and are always changing.

- **High-altitude clouds**: cirrocumulus, cirrus, and cirrostratus; cumulonimbus can form giant clouds and reach very high.
- **Mid-altitude clouds**: altocumulus and altostratus.
- **Low-altitude clouds**: stratus, stratocumulus, cumulus.

Weather: temporary changes in climate
Climate: long-term changes in climate

The six key components in describing weather are **temperature**, **atmospheric pressure**, **wind**, **humidity**, **precipitation**, and **cloudiness**. These **influence the movement of currents** in the atmosphere.

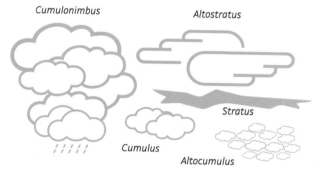

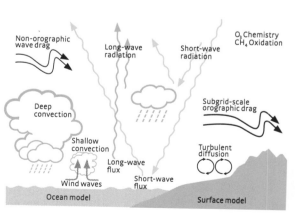

CARBON CYCLE

Carbon is cycled through CO_2 and glucose ($C_6H_{12}O_6$) in photosynthesis and glycolysis (the breakdown of glucose); this interacts with the hydrogen cycle, which includes H_2 and H_2O; this in turn interacts with the oxygen cycle.

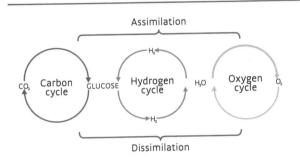

1. Carbon dioxide is released into the atmosphere from **respiration** and **combustion**. Plants release CO_2 **during the night** when they cannot photosynthesize.
2. Carbon dioxide is **photosynthesized to form glucose** in the presence of **sunlight**.
3. Animals feed on plants, **metabolizing** and **digesting** the glucose and **liberating carbon atoms**, most of which are exhaled as carbon dioxide during **aerobic respiration**.
4. Organisms **die**.
5. **Decay** releases carbon atoms into the atmosphere in **carbon dioxide** and **methane**.
6. Over millions of years, **fossils transform into crude oil and gas** due to geophysical processes.

CARBON FIXING

The assimilation of carbon (CO_2) into **carbon compounds** by **biochemical processes**. **Photosynthesis** is an example of this, but plants eventually die and release CO_2 when they **decompose**. The most permanent **carbon-fixing process** takes place in the formation of **limestone** by **coral reefs**.

GREENHOUSE EFFECT

Some gases such as CO_2, methane CH_4, water vapor H_2O, nitrous oxide N_2O, and ozone O_3 trap the Sun's warming infrared rays in the lower atmosphere. These **greenhouse gases absorb and emit thermal energy**. Increasing amounts of CO_2, CH_4, and N_2O from human industry contribute to an increase in thermal energy being retained, resulting in a more energetic climate, which increases rates of evaporation.

THE GREENHOUSE EFFECT

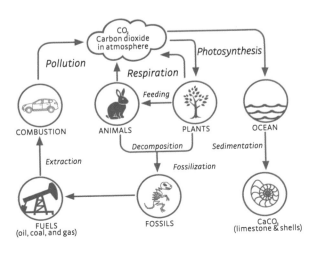

METHANE

Methane (CH_4) is the second most important greenhouse gas. It forms naturally from **decomposition in wetlands**, **organism digestion**, and in **oil and gas formation**. As the **polar tundra** thaws due to **global warming**, more methane is released into the atmosphere.

ROCK CYCLE

The Earth formed 4.5 billion years ago from gravitational processes and the remnants of a supernova. A lot of iron worked its way down to the core. The Earth's core is about 80 percent iron and also contains nickel, gold, platinum, and uranium.

TYPES OF ROCK

- **Igneous**: formed when hot **magma** rapidly cools; **granite**, **obsidian**, and **pumice** are igneous rocks.

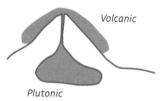

Volcanic

Plutonic

- **Sedimentary**: formed by layers of **sediment** (eroded sand and rock particles) compressed in layers building up over millions of years; **limestone** and **sandstone** are sedimentary and often contain **fossils**.

- **Metamorphic**: from when rocks are **compressed and twisted** together under **intense pressure and heat**; these rocks are very hard, and sometimes contain **crystallized minerals** from cooling very slowly; **marble** and **slate** are metamorphic.

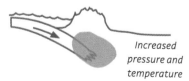

Increased pressure and temperature

THE ROCK CYCLE

1. **Weathering erodes** igneous, sedimentary, and metamorphic rock.
2. **Transportation** of eroded rock particles in **rain**, **streams**, and **rivers** lead to the **sea**.
3. **Deposition** of rock particles in the **ocean sink**, building up **sediment**.
4. **Compaction** and **cementation** of sediment occurs under increasing **weight and pressure**, **compacting** the layers beneath.
5. **Metamorphism** occurs over millions of years: sedimentary or igneous rock is affected by **tectonic movement** and **subducted**, **deformed**, **twisted**, and **compressed**; **high temperature** and **pressure** create metamorphic rock.
6. Metamorphic rock melts into magma; in **volcanic eruptions** or at the **interface of tectonic ridges**, magma flows out and is cooled, feeding back into the system.

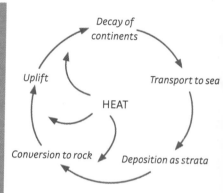

MARINE FOSSILS IN THE HIMALAYAS

The **fossils** of **ammonites**, **shells**, and **other marine life** can be found in **limestone** in the Himalayas. The Himalayas rose from **tectonic movements**.

EROSION AND LAND FORMATION

The **rock cycle** interacts with the **hydrological cycle**; **weathering** and **erosion** shape the landscape. **Glacial erosion** creates **fjords** and **mountains**.

GEOMAGNETISM

The Earth's magnetic field emerges from free-flowing electrons released from liquid iron in the mantle. The rotation of the Earth and convection currents in the mantle give rise to an electromagnetically induced magnetic field.

MAGNETIC FIELD

Characterized by incoming **field lines**, the north magnetic pole does not exactly align with the **geographical North Pole**.

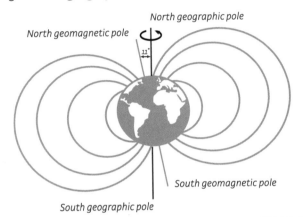

GEOMAGNETIC REVERSAL

Geomagnetic reversal is **when magnetic north and magnetic south swap—reversing magnetic polarity**. There is evidence that this has occurred numerous times in the Earth's history. Evidence of the last reversal, which took place during the **Stone Age**, 780,000 years ago, can be seen in **rock formations**. Geomagnetic reversal is a process that takes about 7,000 years to complete. The Earth's rotation causes curved trajectories and currents in the mantle and outer core; the resulting turbulence twists and shears the old field and regenerates a new one.

MAGNETIC SHIELD

The Earth's magnetic field envelops the Earth in a **protective** shield that deflects harmful high-energy **cosmic rays** and **photons** that would **irradiate** living organisms and **strip away layers of the atmosphere**. The Earth's magnetic field extends for **tens of thousands of miles around it into space**.

AURORA

- **Charged particles** emitted by the **Sun** are sprayed out in all directions from its surface.
- The **solar winds** eventually reach the Earth and are **deflected by its magnetic field**.
- Charged particles are **accelerated** and **collide with molecules** in the upper atmosphere, resulting in **sprays of photons**.
- **Different colors indicate different collisions.**

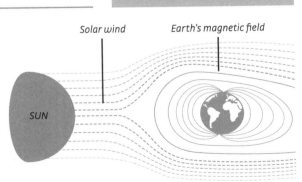

BIOACCUMULATION

Bioaccumulation is the buildup of industrial, agricultural, and toxic chemicals in organisms and ecosystems. Food webs and nutrient cycles are deeply interconnected in the biosphere. Toxic substances (heavy metals, radioisotopes, and other compounds) build up in organisms due to their chemical properties.

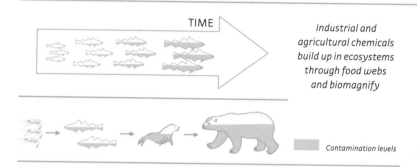

Industrial and agricultural chemicals build up in ecosystems through food webs and biomagnify

Contamination levels

INSECTICIDES AND WEED KILLERS

Insecticides and weed killers have **toxic bioaccumulative effects**.

DDT

DDT (Dichlorodiphenyl-trichloroethane) and related chemicals such as DDE and DDD **are lipophilic** (stick to fat molecules) and **remain in the body and soil for decades**. DDT is accumulated in the body. Over decades it is excreted in urine, feces, and breast milk. Children born decades after the ban have **traces of DDT** in their bloodstream.

SILENT SPRING

Rachel Carson (1907–64) wrote **Silent Spring** in 1962. It explained how **pesticides work their way through food webs, damaging the environment** and **killing wildlife**. DDT was a key culprit. **Birds** were severely affected, DDT **preventing their bodies from making eggshells**.

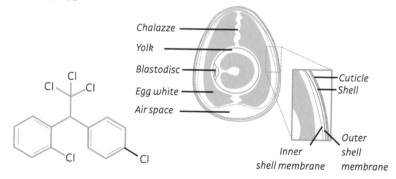

Silent Spring criticized **humans' exploitation of nature** in the name of **so-called progress**. Carson's findings were not new; scientists were aware of the problem but **Carson brought them to public attention**.

STRONTIUM-90 AND BIOACCUMULATION

The bioaccumulative effects of **cancer-causing radioactive nuclei** from **nuclear weapons testing** has also been measured. It continues to circulate food webs.

Thanks to Carson, many countries across the world have **banned DDT**, including **India**, **China** and countries in the **Americas** and **Europe**. **Some equatorial countries** struggle to manage the spread of insect-borne infections such as **malaria** and sometimes **use DDT**. This is a complex issue.

HUMAN-MADE CLIMATE CHANGE

The Earth's climate, biosphere, oceans, and cycles are interlinked. Industrial human activity is interrupting and distorting them, causing the destruction of habitats, inequality, pollution, increasing global temperatures, and the extinction of organisms.

CLIMATE

This denotes **long-term weather patterns**.

Interaction among **atmosphere, ocean**, and **land** causes **natural weather changes**, but **disrupting its balance** causes **sudden and long-term changes in climate**.

ENERGETIC SYSTEM

The Earth's systems are becoming more **energetic** (rather than simply warmer), resulting in **extremes** in **all kinds of weather**.

WATER

The **hydrological cycle** is becoming more active. **Increasing global rainfall, floods, evaporation**, and **snow** are affecting **water supply and quality**.

HOLE IN THE OZONE LAYER

In the 1960s and 1970s, the Western world emitted **CFCs (chlorofluorocarbons)**, which eroded the ozone layer. The **Montreal Protocol** signed in 1987 **banned CFCs**. Fortunately, **the hole in the ozone layer is beginning to heal**.

INSECURITY

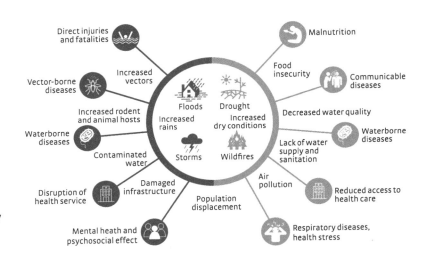

RACISM AND CLIMATE CHANGE

- People living in the **Global North** overwhelmingly **consume more**, and **produce more waste and CO_2 per person**, compared to those living in the **Global South**.
- **Indigenous communities** experienced **historical genocide** at the hands of **Western nations** following the **fifteenth-century invasions of the Americas (and Australia)**. To this day, **indigenous communities have to fight to protect their land from devastation**.
- The failure of mainstream conversations to discuss the fact that the **wealth of the Global North** and **extractive practices** generally are a **direct historical consequence** of **past empires, colonialism,** and **current neocolonial exploitation** means that the **depth and complexity of climate issues are not being confronted or resolved**.
- **Developing economies** that currently **rely on industries that pollute are being blamed for being poor by countries that exploited them to the point of poverty**.
- The **Global South is at the front line of climate change**, experiencing directly the worst devastation.

THE LIBRARY OF ALEXANDRIA

Organizing knowledge is a very important and ancient practice. Librarians to this day work hard to keep our growing body of knowledge (including digital knowledge) organized so that it can be found.

ESTABLISHMENT AND DESTRUCTION OF THE LIBRARY

The library of Alexandria was established about 285–246 BC in **northern Egypt**. It was **one of the largest libraries in the ancient world**. It's said to have contained works on **ancient languages**, **poetry**, **music**, **systems of thinking**, **mathematical knowledge**, and **writing**. The library contained **scrolls** and was **celebrated as a center of learning**.

Some historians claim that the library was **neglected** and that **over time the scrolls were gradually disintegrating**. Others say that the library was destroyed **"by accident"** in about 48 BC in a **huge fire** during a **siege of Alexandria** instructed by **Julius Caesar**. Some say that it was burned down by **religous fundamentalists**. **Remains** of the library **survive to this day**.

HYPATIA

Born about AD 350–370, Hypatia was a **mathematician** and **astronomer**. She developed ideas about **geometry** and **number theory**. Around AD 410 she was **attacked by a mob of extremists** who disagreed with her **Pagan beliefs**. Tragically, she was **murdered**. Hypatia remains a **symbol of knowledge and wisdom in the face of ignorance**.

LIBRARY MANAGEMENT

The management of archives and libraries requires continual maintenance, such as the updating and tracking of items in a library's collection. The same applies to digital libraries. Libraries need to be accessible, well-organized, and up-to-date.

THE MODERN BIBLIOTHECA ALEXANDRINA

Between 1988 and 2002, a **new Bibliotheca Alexandrina** was built. It received funding to include a **digital archive of billions of Web pages, millions of Web sites that no longer exist online**, and **extensive television and audio broadcast archives**.

THE CIRCUMFERENCE OF THE EARTH

Over 2,000 years ago, Greek mathematician Eratosthenes measured the circumference of the Earth using nothing but shadows, trigonometry, and the distance between two cities.

MEASURING THE EARTH'S CIRCUMFERENCE

At noon on the **summer solstice** (June 21), the Sun's rays shone directly down a deep well in the city of Swenett, located near the **Tropic of Cancer**. The city of Swenett was later known as Syene and is modern-day Aswan. At the same time on the same day of the year, the Sun would not shine directly down a well in **Alexandria**, but would cast a shadow into the well. Eratosthenes noticed the **same effect with an obelisk casting shadows** in Alexandria:
- On the summer solstice at noon in Swenett: no shadow;
- On the summer solstice at noon in Alexandria: shadow.

Eratosthenes measured the angle of the shadow.

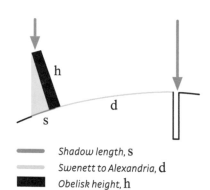

Shadow length, s
Swenett to Alexandria, d
Obelisk height, h

The **distance between Swenett and Alexandria was measured** and, using this distance, the **angle of the shadow**, and some **geometry**, Eratosthenes **calculated the planet's circumference**.
- A circle contains 360 degrees.
- Shadow angle: 7.2 degrees.
- Dividing 360 by 7.2 gives 50.

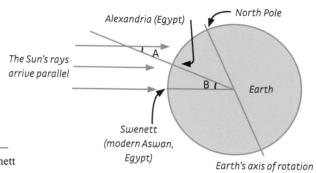

$$\frac{360 \text{ degrees}}{7.2 \text{ degrees}} = \frac{\text{circumference of the Earth}}{\text{distance from Alexandria to Swenett}}$$

Dividing 360° by 7.2° gives 50, which meant that the distance between Alexandria and Swenett (500 miles) was 1/50 of the circumference of Earth.

Eratosthenes multiplied 500 miles by 50 to arrive at his estimate of the Earth's circumference: 25,000 miles. The circumference of the Earth is about 24,901 miles.

Around 270 BC, philosopher **Aristarchus** calculated the **Moon's distance from the Earth**. He assumed that the Moon orbits Earth on a circular path of **radius R in time T**, and observed the time t that **Earth's shadow took to move across the Moon during a lunar eclipse**.

Using some basic geometry, he made the following calculation: it takes time t for **the Moon to move into the segment of lunar orbit** occupied by Earth's shadow, which he estimated as 2r (where r is Earth's own radius).

It takes time T for the Moon to complete an entire orbit of Earth, equivalent to $2\pi R$ or approximately 6.28 R.
$t/T = 1/363 = 2r/6.28 R$
$r/R = 1/60$
Thus the average distance of the Moon is **60 Earth radii.**

THE MEASUREMENT OF TIME

Natural cycles affect human, plant, and animal life and are the basis of ancient calendars and those we use today.

- **Lunar calendars** relate to the **cycles of the Moon**.
- **Solar**, that of the **Sun**.
- **Sidereal**, the **regular changing of the patterns of stars**.

Many calendars incorporate **complex luni-solar and sidereal cycles** even though the **Moon, Sun, and Earth do not relate to one another by exact cycles**.

PHASES OF THE MOON

Lunar calendars are **some of the oldest calendars in existence**.

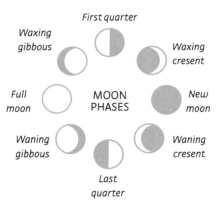

SIDEREAL CALENDARS

Signs of the zodiac are groups of **stars that appear as patterns on Earth**. Over time, these patterns shift due to the **precession of the Earth. Different cultures** all over the world (and throughout history) **have different celestial zodiacs**.

SECONDS, MINUTES, HOURS

A **second** equates to roughly **1/86,400 of a day** or **1/60 of a minute**, which is **1/60 of an hour**. An **hour is about 1/12 of the time between sunrise and sunset**. Except at the equator, **the time between sunrise and sunset changes throughout the year**.

HOW LONG IS A SECOND?

SI (Système International) units are **internationally decided units of measurement**. The definition of the second was finally decided by CIPM in 1967; **it lasts 9,192,631,770 oscillations of a caesium-133 atom**.

STANDARDIZED TIME ZONES

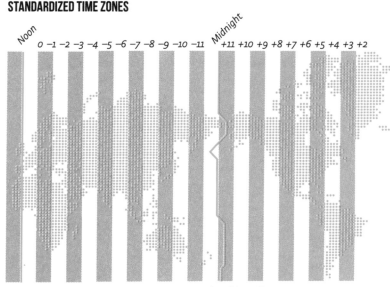

WEIGHTS AND MEASURES

Natural cycles are **not sufficiently predictable for perfectly accurate timekeeping**. In science, this is important. In 1957 the need for a **globally accepted definition of a second** was discussed by the **International Committee of Weights and Measures** (CIPM).

WEIRD TIME

A special property of the universe is that **we can remember the past but not know the future**. This seems an obvious statement but it is **puzzling from a mathematical point of view**.

ISMAIL AL-JAZARI

Ismail al-Jazari (1136–1206) was a Turkish polymath, engineer, and artist of the Islamic Golden Age. He wrote The Book of Ingenious Devices—*published in the year he died. The book illustrated and described over one hundred mechanical devices and automata.*

KITĀB AL-ĀIYAL (*THE TRICKS BOOK* OR *THE BOOK OF INGENIOUS DEVICES*)

An earlier book published in the ninth century by the **Banū Mūsā brothers** at the House of Wisdom in Baghdad included **levers**, **counterbalances**, and **gears**.

DESIGN

Engineers are continuously **designing, prototyping, testing,** and **redesigning** in order to improve the things they make.

AUTOMATA

Ismail al-Jazari's book included **water-dispensing devices**, **clocks**, and automated musical **entertainment**. His devices operated using **gears**, **crankshafts**, **pressure pistons**, **water pumps**, and **water mills**.

ELEPHANT CLOCK

Among Al-Jazari's extraordinary contraptions is an elaborate **water clock** that **counted the hours between sunrise and sunset**. The internal mechanism worked as follows:
- A bowl floating in a water tank slowly filled with water.
- When it sank, it tugged a pulley system, which tilted a ball-bearing-driven mechanism.
- This rotated to tell the time.
- It then dropped a ball into the mouth of a dragon, then into a container next to the elephant driver; the driver then banged a drum to tell the time.

GEAR PRINCIPLES

Gears **translate force** to different mechanisms. They operate by using **ratios of numbers of teeth on a gear** to change **"revolutions per minute,"** or RPM. The **driver** gear is larger than the **driven** gear, such that large to small **multiplies the velocity ratio** of the RPM of the driven gear.

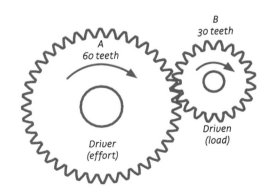

EXAMPLE
60 teeth in A divided by 30 teeth in B = velocity ratio: 60 ÷ 30 = 2
If the RPM of A = 120, then the RPM of B = the RPM of A × velocity ratio: 120 × 2 = 240

MOVABLE TYPE

Printing is the ancient technology of creating and reproducing images and words. The printing press revolutionized the distribution of information. Originally invented in China, movable type made it easier to change words.

TYPES OF PRINTING

 Printing from top of incision is called: **Relief** — Woodcut, linocut, letterpress

 Printing from bottom of incision is called: **Intaglio** — Etching, drypoint, engraving

Printing from a single plane is called: **Planographic** — Monotype, lithography

 Printing through an opening is called: **Stencil** — Silkscreen, pochoir

ANCIENT STENCILS

35,000-year-old **stencils of human hands** have been **discovered on cave walls**. Stenciling is a **type of printing**.

BLOCK PRINTING PROCESS

This used **carved wooden blocks** coated evenly with **ink** and then **pressed onto a surface** such as **paper** or a **woven textile**. There is evidence of printing using **clay tablets** from as early as 3500 BC. **Woodblock techniques** have been used in **China, Japan, Korea**, and **India** for thousands of years.

RELIEF PRINTING

Ink is **applied to a surface**, not reaching areas where **material is carved out**.

INTAGLIO PRINTING

Ink is applied to the areas where material is carved away from the printing plate. The opposite of relief printing. **Etching** is an example.

THE INNOVATION OF MOVABLE TYPE

Movable type **can be adjusted**, making it **easier to change text**. **Individual elements, characters, or letters can be arranged** to change meaning, avoiding the need to painstakingly re-carve or write text from scratch.

BI SHENG

A **Chinese peasant** named **Bi Sheng** (990–1051) developed the world's **first movable type printing system**.

INDIVIDUAL CHARACTERS OR LETTERS

LETTERPRESSING

Typesetters will **compose text back to front**—a **mirror image** of each **letter** or **character** is used to print text the right way round. **Blocks** are used to **space words** from one another and from the edge of the page. A **composed text** is then placed into the **"bed" of a press**, **locked** into place, and **inked**; then **paper is positioned** and **pressed evenly** to create a print.

GUTENBERG PRESS

German blacksmith and metalworker **Johannes Gutenberg** (1400–68) invented the **Gutenberg press** between 1440 and 1450. This **changed how information could be distributed**.

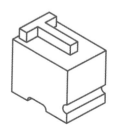

CONSTRUCTION

Civil engineering is an ancient human practice that involved the design and construction of large structures and buildings.

- **Structural engineering**: designing the frameworks needed to construct **buildings and structures; they are needed to ensure buildings are designed to withstand physical stress.**
- **Construction engineering**: involves the planning, management, and construction **of designed structures**, e.g., infrastructure.

CITY PLANNING

The history of city planning **dates back thousands of years**. Planning and developing cities requires decisions regarding **sewers, public transport, water supply, hospitals, transportation**, and **schools**. Many cities **develop and grow over preexisting infrastructure**.

CONCRETE

Concrete is made from a mixture of **powdered stones** and **water**. A **chemical reaction called hydration** between the water and solid powdered ingredients forms chemical bonds between the atoms in the powder, causing them to join together, harden, and solidify like glue. The **crushing and burning of gypsum or limestone is an ancient equivalent**.

LIME

Calcium hydroxide, or **slaked or hydrated lime**, is **caustic** and **causes severe burns**. It is, however, a **versatile construction material**. Used in **masonry** as **mortar**, it **sticks rocks and bricks together** and can be used to **coat surfaces**. Exterior (**stucco**) and **interior plaster** contain lime.

REINFORCED CONCRETE

Concrete is **brittle** and has **low tensile strength**. Reinforcing it with **steel frameworks, rods, bars**, or a **mesh** enables the material to **absorb tensile, shear**, and **compressive pressure**. Reinforced concrete has **enabled the construction of very tall buildings**.

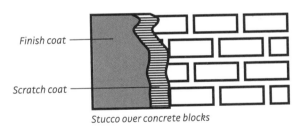

Finish coat
Scratch coat
Stucco over concrete blocks

PORTLAND CEMENT

One of the **basic ingredients of concrete** is **cement**. It contains **fine and coarse particles of crushed calcium compounds, silica, alumina**, and **iron oxide**. Other ingredients include **limestone, sandstone, marl, shale, iron, clay**, and **fly ash**.

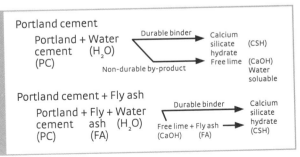

HEAT ENGINE

A system that converts heat into different forms of energy. Heat flows from high to low temperatures and drives (e.g., a turbine), thus converting into mechanical energy. This could be connected to a dynamo that generates electrical energy.

THE HEAT CYCLE

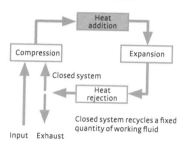

Thermal efficiency = work ÷ heat

FIRST LAW

Energy cannot be created or destroyed; it is **converted** into different forms.

SECOND LAW

It is **impossible to create a 100 percent efficient engine. Some energy** can never be used to do work—i.e., it is **lost as heat**.

SADI CARNOT (1796–1832)

The French engineer behind the **hypothetical Carnot cycle**, which **assumes** the **use of fluids in energy exchange** is **100 percent efficient**. The Carnot cycle is impossible, but **helps engineers understand the limitations of efficiency**.

MOHAMMED BAH ABBA (1964–2010)

This **Nigerian educator** popularized the ancient **zeer pot refrigeration system**. It **uses no electricity** and **keeps food cool in hot climates**. As **water evaporates from moist sand, heat is removed** from the chilled zone containing food.

INTERNAL COMBUSTION ENGINE

Motion is created by **burning fuel with air in a closed system**. The **thermal expansion of gases drives pistons**. Most internal combustion engines in cars have a **low thermal efficiency** of 20 percent.

PHASE CYCLE

The phase transitions of boiling liquids is the driving force behind engines. The two graphs to the right show the **change in phase of a substance in a cycle from liquid to gas**. The **phase cycle** presented is the **Rankine cycle**, which is **used to predict the efficiency of a heat turbine**.

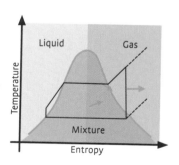

REFRIGERATION

The food inside a fridge stays cold because **heat is continuously being removed from the internal environment. Fluid spends more time as a gas in this system.**

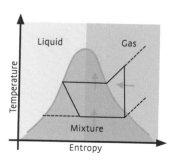

ENERGY FUTURE

Most **electricity** is made from **burning fossil fuels**. Heat generated by **burning evaporates steam**, which **turns turbines**. A **shift in fossil fuel dependence is necessary**.

ENERGY STORAGE

A battery stores electrical potential. The difference in electronegativity provides the potential for a current to flow from negative (anode) to positive (cathode).

Batteries are electrochemical cells made of:
- a **cathode**: a **positive electrode** capable of **gaining electrons**.
- an **anode**: a **negative electrode** capable of **providing electrons**.
- an **electrolyte**: a **fluid that allows ions to flow**, thus creating an **electric current**.

LEMON BATTERY

- Copper wire = cathode
- Zinc-coated nail = anode
- Lemon = electrolyte

BAGHDAD BATTERY

An **early battery design** discovered in **Baghdad** dating from 2,000 years ago was made from a **clay pot containing an iron rod encased in a copper cylinder**.

VOLTAIC PILE

Italian chemist **Alessandro Volta** (1745–1827) stacked zinc and copper discs separated by blotting paper soaked in **brine** (salt water). The **zinc and copper are the electrodes**. It worked, but the **salt water eventually eroded the metal**.

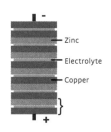

CHEMICAL CELLS

Chemical reactions of **oxidation** and **reduction** inside a battery **cause ions to flow**. **Zinc-carbon batteries** were **invented by** French chemist **Georges Leclanché**.

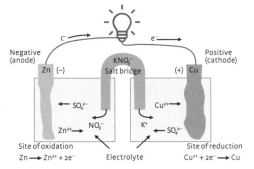

DRY BATTERY CELL

Dry cells developed in 1886 by German scientist **Carl Gassner** used a **paste instead of a liquid**. **Modern batteries** were **first developed by Japanese watchmaker and inventor Yai Sakizo** in 1887.

RECHARGEABLE

Alkaline batteries can **only be used once**. Developing chemical reactions to be **reversible** made rechargeable batteries possible. The **first rechargeable lead acid battery** was **invented** in 1859 **by Gaston Planté**. Modern **lithium-ion batteries** are used in **laptops** and **phones**.

CHEMICAL WASTE

Batteries cause **horrendous waste**.
- **Rechargeable batteries** have decreased dependence on single-use batteries but still **contain toxic materials** that **need to be mined**, and are **dangerous to handle**.
- Devices do not last long—adding to electronic waste.

GREEN ENERGY

Electricity can be made using **wind, sun**, and **tides**. **Storing this energy is an urgent problem in design and engineering**.

THE COMPUTER

Computers are designed to calculate arithmetic or perform logical operations. Computer programming makes the computer environment adaptable.

ANATOMY OF A COMPUTER

- **Hardware**: can be seen and touched; e.g., **keyboard**, **monitor**, and **mouse**.
- **Software**: **instructions** that make the computer **carry out tasks**.
- **Input**: user **enters information**; e.g., words, numbers, sound, and pictures.
- **System/processing**: computer carries out calculations that interact with storage devices and communication networks.
- **Output**: results presented through tactile devices, visual or auditory information.

THE JACQUARD LOOM

The Jacquard loom is the **ancestor of computers**, designed in 1804 by **Joseph Marie Jacquard** to **mechanize the weaving of intricate fabrics**.

THE ANALYTICAL ENGINE

This was a proposed mechanical general-purpose computer designed by English mathematician and inventor **Charles Babbage** (1791–1871) to carry out **complex mathematical calculations**.

SIMPLE COMPUTER ARCHITECTURE

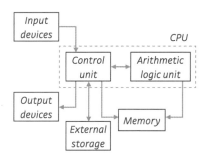

THE ENCHANTRESS OF NUMBERS

English mathematician, translator, and writer **Ada Lovelace** (1815–52) **developed codes for Babbage's computer** that represented **numbers**, **letters**, and **symbols**, and developed a method for the machine to **repeat instructions in loops**. Her work was largely forgotten until the 1950s; loops are **used in computing today**.

BRIEF COMPUTER HISTORY

Generation 1:
- → **No operating system**
- → **Switches** used as **binary codes**
- 1937 Electronic digital computer built
- 1943 Colossus built for military use during the Second World War
- 1946 Electronic Numerical Integrator and Computer (ENIAC) built; performed one mathematical task

Generation 2:
- → **Transistors** used **instead of vacuum tubes**
- → **Computer programming languages** developed
- → **Computer memory** and **operating systems**
- → **External memory**; e.g., tape, punch card, and disks used

- 1951 Universal Automatic Computer (UNIVAC 1)
- 1953 International Business Machine (IBM) made

Generation 3: 1963–present
- → **Integrated circuits**
- → Computers became **smaller, more powerful, and reliable**
- → Use **different programs** at once
- 1980 Microsoft Disk Operating System (MS-DOS)
- 1981 IBM introduced the personal computer (PC)
- 1984 Apple released Macintosh computer and icon interface
- 1990s Windows operating system
- 1992 Smartphones
- 2007 Commercial smartphones released

ELECTRONICS

Electronics is the physics of the flow of electrons—tiny subatomic particles with negative charge that can be harnessed by a huge range of technologies.

KIRCHHOFF'S FIRST LAW

Current flowing in equals current flowing out (**charge conservation**).

KIRCHHOFF'S SECOND LAW

The sum of all voltages in a closed loop equals zero (**conservation of energy**).

PIEZOELECTRIC ENERGY

Piezoelectric material (like quartz) **creates an electric current when it is compressed**. A **quartz watch** contains piezoelectric components within which an **electric current can oscillate with precision and keep time**.

ELECTRONIC SYMBOLS

Variable resistor	Battery	AC Supply
Cell	Diode	Voltage resistor
Resistor	Transformer	Light
Variable resistor	Voltmeter	Solar cell
Ammeter	Switch	Multi switch
Capacitor	Earth	Motor

TOTAL RESISTANCE

(A) RESISTORS IN SERIES:

Total resistance = sum of individual resistors added together.

$$R_T = R_1 + R_2 + R_3$$

Resistors connected in series

(B) RESISTORS IN PARALLEL:

Reciprocal (1/R) value of resistances adds together to give 1/R total.

$$\frac{1}{R_T} = \frac{1}{R_1} + \frac{1}{R_2} + \frac{1}{R_3} \ldots \frac{1}{R_n} \text{ etc.}$$

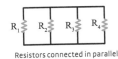

Resistors connected in parallel

READING RESISTORS

Resistors carry **color-coded bands** to indicate their resistance in **Ohms Ω**. Two or three bands on the left indicate digits 0–9, followed closely by a band indicating **multiplying factor**. After a gap, another band indicates **percentage tolerance**.

For example, a **resistor code**:
Yellow violet red … silver
= 4   7   x100   10%
= 4700 Ohms with a tolerance of ± 10 percent.

OHMIC AND NON-OHMIC RESISTORS

Ohmic: **resistance** is **constant**, **current** is **proportional** to potential difference, and **Ohm's law** can be applied. V plotted against I is **linear**: V = IR.

Non-Ohmic: **non-linear** plot for V against I.

- **Filament lamp**: as the filament gets hotter, resistance increases.
- **NTC thermistor**: resistance falls with increasing temperature.
- **Semiconductor diode**: in one direction, current is almost 0; in the other, current increases.
- **Light-emitting diode (LED)**: in one direction it conducts and emits light, but not in the reverse.
- **Light-dependent resistor (LDR)**: resistance changes with light intensity.

THE RESISTOR COLOR-CODE TABLE

Color	Digit	Multiplier	Tolerance
Black	0	1	
Brown	1	10	± 1 percent
Red	2	100	± 2 percent
Orange	3	1,000	
Yellow	4	10,000	
Green	5	100,000	± 0.5 percent
Blue	6	1,000,000	± 0.25 percent
Violet	7	10,000,000	± 0.1 percent
Gray	8		± 0.05 percent
White	9		
Gold		0.1	± 5 percent
Silver		0.01	± 10 percent
None			± 20 percent

TECHNOLOGY

ALAN TURING

English computer scientist, cryptanalyst, philosopher, theoretical biologist, and mathematician Alan Turing (1912–54) developed the modern computer and AI. During the Second World War he worked at Bletchley Park deciphering Nazi codes sent using the Enigma machine—thus helping to defeat them.

ENIGMA

Enigma machines made use of **cogs** and **electronics** to **"randomly" scramble messages**, but machines have a **limited number of operations** they can carry out. **Turing figured out their mechanism.**

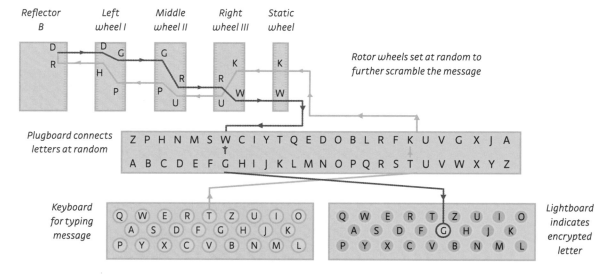

TURING'S TRIALS

At **Cambridge in the 1930s**, Turing worked on **mathematical problems** around the limits of **computing machines**. His idea of the **"Turing machine"**—a hypothetical device that simply processes numbers according to a set of rules, and yet is capable of solving any problem that can be written down as an algorithm—was highly influential.

He was **recruited by government codebreakers** in 1938 to help find **weaknesses in the German Enigma code**, and soon devised electromechanical machines called **bombes** that could use brute-force calculating power to sift through possible Enigma settings (based on a guessed-at fragment of the original message called a "crib"). His work influenced **Colossus**, the world's first programmable digital computer, built at **Bletchley**.

After 1945, however, work from Bletchley Park was **classified** and the machines broken up, hampering Turing's postwar efforts to build a more advanced **"stored-program"** computer. Furthermore, Turing's **homosexuality** was deemed a security risk and led to him being **excluded from intelligence work**. After arrest in 1952 under the homophobic laws of the time, he was ordered to receive a form of **hormonal castration**.

A combination of **frustration and depression** drove him to suicide in 1954. His **huge contribution** to the modern world has only been recognized in recent decades.

PHOTOGRAPHY

The invention of photography involved photoreactive chemistry, optics, and visual art. Photography was rapidly adopted by both artists and scientists.

THE PINHOLE CAMERA

- Can be made without a lens.
- Makes use of a tiny aperture (hole).
- Light enters the aperture, projecting an inverted image inside the box.

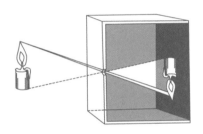

CYANOTYPE

An early form of **photoreactive image making** that uses a **coating applied to sheets of paper**. When **exposed to sunlight**, parts **covered by an object** are **imprinted onto the paper**.

ANNA ATKINS

English botanist **Anna Atkins** (1799–1871) was **one of the first female photographers** and published the **first book of photography featuring cyanotype images of plants**.

DAGUERREOTYPE

Louis-Jacques-Mandé Daguerre (1787–1851) invented the daguerreotype process using **copper plates coated with silver**.

DIGITAL

Light enters the camera lens, which **stimulates an image-sensor chip**, which **measures color**, **tone**, and the **contours of different shapes**. This **analog information** is **translated into millions of pixels**.

LENSES

Lenses can be used to **project images onto canvases and other surfaces**. This enabled Flemish artists of the early Renaissance to trace objects and people, and create realistic paintings.

ANATOMY OF AN SLR CAMERA

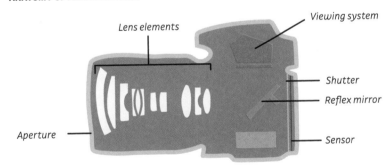

Labels: Lens elements, Viewing system, Shutter, Reflex mirror, Sensor, Aperture

CELLULOID FILM

John Carbutt, **Hannibal Goodwin**, and **George Eastman** developed **transparent flexible film** made from **nitrocellulose, camphor, alcohol**, and **pigments**. Eastman Kodak made **celluloid film commercially available** in 1889.

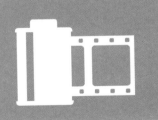

RADAR AND SONAR

Radar and sonar locate objects using the echoes of acoustic signals received from a distant object. Time delays between sending and receiving a signal make location possible.

RADAR

Uses **radio waves**. Among other things, radar **enables planes to land safely** in the dark and through clouds.

SONAR

Uses **sound waves underwater or through the human body**. Modern **ultrasonic generators** emit frequencies between 20,000 Hertz (Hz) and 1 gigaHertz (1 GHz = 1 billion Hz).

ECHOLOCATION

Used by **whales, shrews, dolphins, bats**, and other echolocating animals. Echolocating animals will **emit calls** and **detect echoes** of those calls in order to **sense their environment**, **locate objects**, and **detect distances**. The **distance between sound receptors**—e.g., ears—enables **precise measurement**.

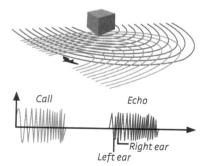

HIGH-DEFINITION HEARING

Bats emit very **loud high-pitched calls**. The **frequency** of their calls is **beyond human hearing**, but is **louder than a jumbo jet at take-off**. Their high-frequency calls have a **small wavelength** and enable them to **detect insects in the night**. A high frequency also enables them to **detect tiny features on an insect like a moth**; e.g., if it has a furry body or antennae. This high-definition detection **works only at short distances**. Bats can **muffle their ears** so that they are not deafened by their own calls.

NOISY OCEAN

Our oceans are becoming **increasingly busy and loud environments** that **disrupt whale communication**. Whales depend on their **song** to **echolocate**, **hunt**, and **communicate** with one another.

A whale sends out its sound and songs...

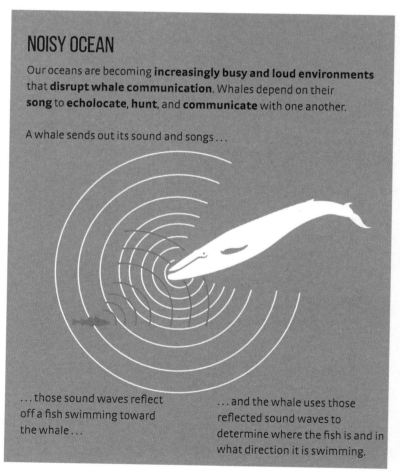

...those sound waves reflect off a fish swimming toward the whale...

...and the whale uses those reflected sound waves to determine where the fish is and in what direction it is swimming.

INFORMATION

Data from sensors needs to be digitized to be processed by a computer. Data used by computers comes in the form of electrical potential difference.

ANALOG

A **signal from a continuous potential difference** (voltage) **cannot be stored or processed directly in a computer**. Information varies, depending on **amplitude** and **wave shape**, and must be **digitized** before being processed by a computer. **Analog waves** come in many forms.

DIGITAL

In digital systems a signal is formed from **two values (0 or 1)** that are decided using two voltage thresholds that allocate signals to 0 or 1. A signal formed from **0 or 1** is called a **bit**. Eight bits equals one byte, and the **resolution** depends on the number of bits. Digital signals are **easy to recover** if conditions cause **noise** or **interference**, as a signal **doesn't depend on amplitude or wave shape but on the sequence of pulses**. To **convert analog to digital**, a **sine wave is converted into a square wave or combination of square waves**.

BANDWIDTH

Bandwidths vary depending on the quantity of information that needs to be transmitted. **A higher rate of transmission requires a broader bandwidth.**

WAVEFORMS

Electromagnetic variations are used to transmit information. **Metal wires or cables** can be used; variations are transmitted **atmospherically** in the case of **radio**; or via **photons** in **fiber optic cables**.

Attenuation: the energy of a signal dissipates over distance.

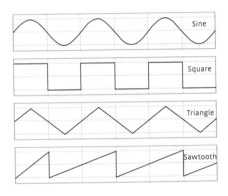
Sine / Square / Triangle / Sawtooth

INTERFERENCE AND SIGNAL DEGRADATION

This can be caused by **thermal radiation** produced by **friction**; **reflective feedback** in fiber optics; or **atmospheric interference** in the case of radio and microwaves.

FREQUENCY SPECTRUM

To transmit sound electromagnetically, audible frequencies must be sampled at 40 kHz at a resolution of 16 bits. Each second transmits a rate of 640 kHz—640,000 bits per second.

Systems that transmit electromagnetically:
- Radio
- Cellular communication
- Wi-Fi

CARRIER FREQUENCY

Information for transmissions is encoded using the **modulation of a carrier wave**. The carrier wave contains the energy (expressed as a frequency) needed to transport information.

MODULATION

Modulating a signal refers to **changing a signal over time**. Modulation requires that one or more properties of a periodic waveform (or carrier signal/frequency) is modified to carry information. This may require **frequency (FM) or amplitude modulation (AM)**.

SOUNDS

The human-audible sound range is between 50 Hz and 15 kHz.

GPS

Global Positioning System (GPS) is a navigational system comprising about thirty satellites orbiting at an altitude of about 12,500 miles around the Earth. Working on a basis of trilateration, whereby information on geographical locations is gathered from at least three different satellites, allows location to be precisely charted using GPS.

At the dawn of the twentieth century, **faster travel by train**, and electrical communication **by telegram**, meant that **standardizing time zones was essential** in order to avoid train collisions. Time zones were introduced to help manage faster travel and communication. After the **duration of a second** became fixed with a **theoretical definition** across the world, **synchronizing activities** became possible.

HOW IT WORKS

GPS works by **comparing satellite data and sending electromagnetic signals**:
- Signal is sent from GPS to the nearest satellite.
- Signals are then sent to at least three other satellites.
- Distance of each satellite is determined using **time-delay data** which is sent to GPS device.
- **Comparison between the position and distance of at least three satellites** reduces the error rate and **accounts for movement**.
- Using this information, the receiver **calculates the device's position**.

GLADYS MAE WEST

African-American mathematician **Gladys Mae West** (born 1930) contributed significantly to the development of the **mathematical models of the Earth** essential to the operation of GPS. Her vital work contributed to the **satellite geodesy models** that were incorporated as part of the GPS used today. Her role was to **measure the positions of satellites and calculate their orbits** using the computer software of around 1956 to 1960.

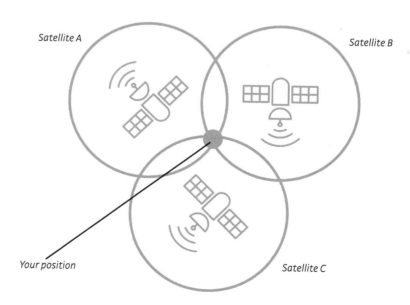

SPACE TRAVEL

More than 500 human beings have traveled beyond the stratosphere and into orbit around our planet. Twelve have walked on the Moon, but so far only robot probes have traveled to other planets.

SPACE TRAVEL MILESTONES

1957	First satellite in orbit (Sputnik 1)
	First animal in orbit: Laika the dog (Sputnik 2)
1958	First US satellite (Explorer 1)
1961	First hominid in space (suborbital): Ham the chimpanzee
	First man in orbit: Yuri Gagarin
	First American in space (suborbital): Alan Shepard
1962	First American in orbit: John Glenn
1963	First woman in orbit: Valentina Tereshkova
1965	First space walk: Alexei Leonov
1968	First crewed spaceflight around the Moon (Apollo 8)
1969	First astronauts on the Moon: Neil Armstrong and Buzz Aldrin (Apollo 11), with Michael Collins remaining in lunar orbit
1971	First space station (Salyut 1)
1972	Final Apollo lunar mission (Apollo 17)
1981	First space shuttle flight (Columbia)
1983	First American woman in orbit: Sally Ride
1986	Assembly of first modular space station (Mir) begins
1994	First visit of space shuttle (Discovery) to Mir
1998	Assembly of International Space Station (ISS) begins
2001	First space tourist, Dennis Tito, visits the ISS
2004	Spaceship One completes first commercial crewed spaceflight
2011	Final space shuttle flight (Atlantis)
	Completion of the ISS

VISITING OTHER WORLDS

These spacecraft were the first to successfully visit other worlds in our solar system:

Moon	Flyby: Luna 3 (1958)
	Soft landing: Luna 9 (1965)
Mercury	Flyby: Mariner 10 (1974)
	Orbiter: MESSENGER (2011)
Venus	Flyby: Mariner 2 (1962)
	Landing: Venera 8 (1972)
	Orbiter: Venera 9 (1975)
Mars	Flyby: Mariner 4 (1965)
	Orbiter: Mariner 9 (1971)
	Landing: Viking 1 (1976)
	Rover: Mars Pathfinder (1997)
Ceres (largest asteroid)	Orbiter: Dawn (2015)
Jupiter	Flyby: Pioneer 10 (1973)
	Orbiter: Galileo (1995)
Saturn	Flyby: Pioneer 11 (1979)
	Orbiter: Cassini (2004)
Uranus	Flyby: Voyager 2 (1986)
Neptune	Flyby: Voyager 2 (1989)
Pluto	Flyby: New Horizons (2015)
Comet Halley	Close approach: Giotto (1986)

HIDDEN FIGURES

Human and robot spaceflight are possible only through the work of thousands of scientists, engineers, and contractors, many of whom remain anonymous. In 2016 Margot Lee Shetterley's book *Hidden Figures* highlighted the contribution of African-American women such as mathematicians Katherine Johnson and Dorothy Vaughan and engineer Mary Jackson to the early days of NASA's space program.

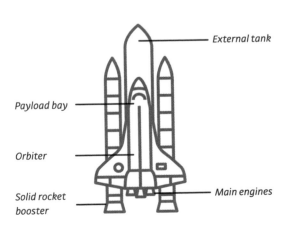

TECHNOLOGY

PROGRAMMING

With the development of programming languages, computers became more versatile.

MORSE CODE

Code is simply a **character substituted for dots and dashes**. Morse code is a good example of a code.

A	·–	J	·–––	S	···
B	–···	K	–·–	T	–
C	–·–·	L	·–··	U	··–
D	–··	M	––	V	···–
E	·	N	–·	W	·––
F	··–·	O	–––	X	–··–
G	––·	P	·––·	Y	–·––
H	····	Q	––·–	Z	––··
I	··	R	·–·		

LEVELS OF COMPLEXITY

The **Anglophone alphabet** has **phonetic sounds**, **grammar**, and **twenty-six letters**. Morse code is composed only of **dots and dashes. Both can convey the same information.**

MACHINE LANGUAGE

Computer hardware processes only **binary instructions**.

THE EARLY DAYS

Instructions were **written from scratch using machine language**, starting with **instructions written on paper** in a **human language**, then **translated into binary**.

ASSEMBLY INSTRUCTIONS

These map instructions to machine operation. Historically, this enabled developments from **linking functions** and **fetching data from memory**, to other **instructions** and **loops**.

COMPILERS

Compilers **transform source code into low-level languages** such as assembly instructions.

FORTRAN, 1954

"**Formula Translation**" compilers made programming easier, but **upgrades required the whole system to be rewritten**.

COBOL, 1959

Through COBOL, computers could now **accept the same source code**, making **updates easier**. Write once—run anywhere!

A BRIEF HISTORY OF COMPUTER LANGUAGES

1960s ALGO / LISP / BASIC
1970s PASCAL / C / SMALL TALK
1980s C++ / Objective-C / Pearl
1990s Python / Ruby / JAVA
2000s SWIFT / C# / GO / Ubuntu

STATEMENTS

Syntax governs the **structure of statements in code**.

ASSIGNMENT STATEMENT

A **series of assignment statements** are **needed in a program**. Assignments must be **initialized** to **set their initial values**. Variables are assigned to values; e.g., b = 7.

CONTROL FLOW STATEMENTS

These are **conditional statements**. **IF** and **WHILE** statements are the **most common**.

FUNCTIONS

Also known as **methods** or **subroutines**. Naming control statements, functions end in "**RETURN.**"

KERNEL

The **core of a computer's operating system**. It controls everything within the system.

ASCII

American Standard Code for Information Interchange. ASCII represents alphabetic, numeric, and other characters using a seven-bit binary number.

BUCKMINSTER FULLER

Architect, engineer, and designer Richard Buckminster Fuller (1895–1983) published more than thirty books, popularizing terms such as "Spaceship Earth," "Dymaxion," "synergetic," and "tensegrity."

Fuller aimed to create designs that did more with less, his objective being to **help society with issues such as housing**. He was concerned with the **scarcity of resources** on planet Earth and **global climate change**.

SPACESHIP EARTH

A concept that helps to communicate that the Earth has **limited resources** that need to be considered in **social, economic**, and **design systems**.

DYMAXION HOME

The Dymaxion home was **inexpensive** and easy to **mass produce** on a large scale. It never caught on commercially. Key to its design was its ease of transport and pop-up assembly.

DYMAXION MAP

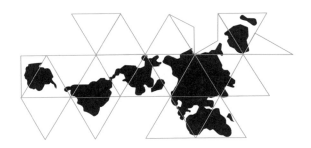

DYMAXION DEPLOYMENT UNITS

These were used during the Second World War to shelter RADAR (Radio Detection and Ranging) crews in remote locations.

THE DYMAXION MAP

Buckminster Fuller designed a **map of the Earth made of triangles** that would fit together to form a **regular polygon**, where **none of the continents or islands were divided or cut**.

GEODESIC DOME

An easy-to-assemble structure that encloses a **large volume of space without intrusive internal supports**. The **stress and mass distribution** within a geodesic dome gave it **remarkable strength**. The geodesic dome is used for **emergency shelters** as well as **research and leisure facilities** across the world.

INSPIRED BY NATURE

Fuller was inspired by observing **geometry in nature**.

SYSTEMS APPROACH PROBLEM SOLVING

Using systems to define **problems** helps us to consider their **interrelatedness** and the possibility of **systems that are in themselves solutions**.

SOCIAL DESIGN

Buckminster Fuller's contemporaries, such as **Victor Papanek**, were similarly concerned with the **effect that design could have in improving society** rather than good design being exclusively for **those who could afford it**.

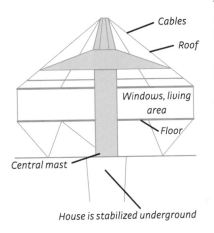

- Cables
- Roof
- Windows, living area
- Floor
- Central mast
- House is stabilized underground

TECHNOLOGY

MAGNETIC RESONANCE IMAGING

Our bodies are made of different substances and materials: proteins, fats, biominerals, neurotransmitters, and water. These materials contain hydrogen atoms, which interact in inter- and intra-molecular bonds. A hydrogen nucleus is a proton.

Protons are key to **Magnetic Resonance Imaging (MRI)**. Protons have **spin**; all the protons in our bodies spin in **different directions** depending on what matter surrounds them. MRI works by **interacting with spin**.

THE BASICS

- A powerful **magnet** in the MRI scanner **aligns the spins** of protons in the same direction.
- **Radio-frequency pulses** (which have less energy than sunlight) then **energize the protons**, moving them **slightly away from their alignment in the magnetic field**, depending on what material they are in; e.g., **protein, fat, bone**.
- The protons are then **refocused by the magnet** and **reemit the energy they absorbed**.
- This is detected by a **radio-frequency coil** in the scanner.
- The **frequencies of energy** emitted by protons vary. The **variation in their signal** is then **processed by a computer to create an image**.

FOURIER TRANSFORMS

A **type of mathematics** used in **analyzing waveforms**. A **signal** such as the sound of a violin is **made of many different frequencies**, and a Fourier transform **will pick up all the different frequencies that create it**. The frequencies emitted by the protons in the MRI scanner are **recombined using Fourier transform calculations**.

BASIC FOURIER SERIES

The basic Fourier series below **shows how to represent a periodic function as the sum of multiple individual (discrete) exponential functions**.

		F (Hz)	A (m)
∿∿∿	∿∿∿	1	1
∿∿∿	∿∿∿	3	1/3
∿∿∿	⊓⊔⊓⊔	5	1/5
∿∿∿	⊓⊔⊓⊔	7	1/7
∿∿∿	⊓⊔⊓⊔	9	1/9
	⊓⊔⊓⊔	11	1/11
	⊓⊔⊓⊔	13	1/13
	⊓⊔⊓⊔	15	1/15
	⊓⊔⊓⊔	17	1/17

DIAGNOSIS, THERAPY, AND MONITORING

MRI scanners are used for **diagnosis, therapy**, and **monitoring** in various **clinical (health care) and experimental (research) contexts**.

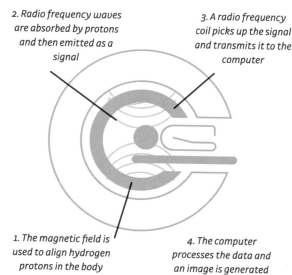

1. The magnetic field is used to align hydrogen protons in the body
2. Radio frequency waves are absorbed by protons and then emitted as a signal
3. A radio frequency coil picks up the signal and transmits it to the computer
4. The computer processes the data and an image is generated

TECHNOLOGY

THE INTERNET

The internet connects networks of computers and devices. Tim Berners-Lee invented hyperlinks (and therefore the World Wide Web) in 1989 as a way for scientists at CERN to share information. It changed the world forever.

HTML

Web pages are built using a **programming language** called **Hypertext Markup Language (HTML)**. HTML embeds the following types of data:
- **Information** included on the page;
- The **design and layout** of the page (formatting);
- **Links** to other pages/sites.

HTML text must be saved as a "**.html**" file.

EXAMPLE:
```
<html>
   <body>
      <h1>Hello world</h1>
      <p>This is a Web page</p>
   </body>
</html>
```

Tags describe layout:
<html> indicates that the text is an HTML document;
<body> identifies the information on the page;
<h1> identifies a heading;
<p> begins a paragraph.

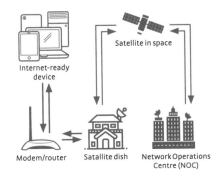

INTERNET OF THINGS

Smart devices in the home connect to the internet. **Malware** is a threat to the internet of things.

TERMINOLOGY

- **Data packets**: Information about the origin and destination of data between computers is broken down into bits called "packets."
- **IP address**: Internet Protocol—a unique number representing the address for a computer.
- **Switch or hub**: Connects devices.
- **Router**: Directs information around the internet.
- **DNS**: Domain Name System turns a Web site into an IP address; it is the protocol that computers use to exchange data.

INTERNET ORIGINS

The internet began life simply as a **means of connecting computers**, sponsored by the **US Advanced Research Projects Agency (ARPA)** in the late 1960s. The first message was sent between computers in 1969, with the ability to send files between computers added in 1971, the same year in which the **first simple e-mails were sent**. However, the modern democratized "World Wide Web" has became a reality only with the **invention of the HTML language in 1989**. Today, faster communications, smaller computers, and wireless networks have made the internet an intrinsic part of our lives, with both benefits and drawbacks.

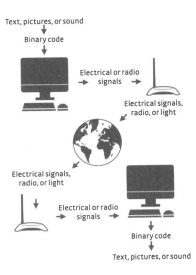

GENETIC ENGINEERING

Genetic engineering involves changing DNA to affect the nature of living organisms. This is controversial, but done ethically it can produce more food, cure diseases, and lead to the invention of new materials.

DNA

Genetic engineers **swap or edit genomes**.

Genes give an organism its **characteristics**; e.g., lions, tigers, and jaguars have different markings.

TRANSGENIC ORGANISMS

These are **organisms with edited genomes**. Golden Rice is one such example: a strain of rice was bred to have **vitamin A** in it to **address vitamin-deficiency-related diseases**.

DIATOMS

Algae with a **silica cell wall**. Their genes can be **edited to create sensors** or act as **drug-delivery units**.

CYTOCHROME P450

Some **bacteria** and **plants** produce **anticancer enzymes**, e.g., cytochrome P450. They can be **genetically modified to produce more of this enzyme**. Insulin used to manage **type 1 diabetes** is made using **gene editing**.

CRISPR AND BACTERIA

Clustered Regularly Interspaced Short Palindromic Repeats (CRISPR) is a **gene-editing technique** that makes use of **bacterial defense mechanisms**. Bacteria produce the **CAS9 protein** to "remember" viral infections **by chopping out sections of viral DNA and RNA**. Treatments for **cancer** and **sickle cell anemia** using CRISPR are being developed.

DESIGNER BABIES

Deleting or **editing** the **genes in an embryo is contested ethically**. Deleting genes that cause known diseases in an unborn human may have benefits but the idea of gene editing to erase the plurality of atypical human bodies and minds, and **selecting traits** such as hair color and deselecting (or deleting) diversities and disabilities has alarming similarities to discriminatory ideas of the past.

CROP FARMING AND GMO

When organisms **reproduce**, some or all of their genes are **passed on to the next generation**. For millennia, humans have **selectively cultivated crops** to **improve yield**, and **resist disease and drought**. Selective breeding is the **same as selecting genes**.

BACTERIAL PLASMID DNA EDITING

Bacterial plasmid DNA is commonly used in bacterial cloning. A specific site on the genome will be used for this.

3-D PRINTING

3-D printing is a process of "additive making" where a substance is liquidized or cured in layers to build up a solid or hollow 3-D structure.

MATERIALIZING DATA

3-D data is used to create an object. This offers creative opportunities for visualizing data and for drawing digital worlds into physical existence. Different materials can be used: **clay**, **cement**, **resin**, **powdered sugar**, and a wide range of **plastics**. Increasingly sophisticated models are being made using 3-D software and ever more materials are being experimented with by **designers**, **artists**, and **scientists** alike.

The capacity for **customization** and developments in the printing of **articulated models** has made 3-D printing a useful technology for **more effective prosthetics**.

LIMITATIONS

3-D printing promised to **democratize manufacturing**, but, as with many innovations, **few people have access to it**. It is also not a substitute for learning how to manipulate a material such as wood, fibers, or clay through one's body and senses.

DRUG DEVELOPMENT

Chemists have developed methods of using 3-D printing to **custom-print medicines**.

BIOPRINTING

The printing of **biologically compatible tissues**. Famously, a **steak has been 3-D printed from stem cells**. Stem cells have been used in 3-D printing to create **cartilage** and **bone**, and engineers hope to move on from this to develop **noses** and **ears**, which could benefit **burn victims**, among others.

ORGANS

The body and **immune system** can **reject a transplanted organ**. 3-D printing may be able to address organ rejection using **pluripotent embryonic stem cells** and a patient's **adult stem cells**. The process is problematic due to the **delicacy of embryonic stem cells**.

METHODS OF ADDITIVE MAKING BY MEANS OF 3-D PRINTING

Cartesian: Based on the Cartesian coordinates of x, y, and z.

Delta: Nozzle and printhead drawing each layer of the object.

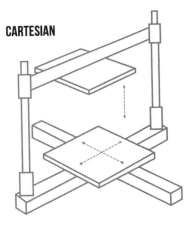

CARTESIAN

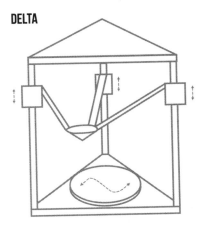

DELTA

TECHNOLOGY

TOUCH SCREENS

Quantum mechanical particles can pass through barriers without going over them—a process called quantum tunneling that underlies modern touch-screen technology.

If we kick a ball at a wall, it will bounce off. In quantum mechanics, rather than acting like footballs, **particles can probabilistically appear in a number of places**. The particle might suddenly appear beyond a barrier, **simply because it can** possibly be on the other side of it. This effect is called **tunneling** and has been **observed in nature and in laboratories** across the world.

SUNSHINE

The reason the sun shines is because **protons tunnel through an energy barrier**. We are familiar with the notion that **like charges repel**, but **because protons are quantum mechanical in size and can tunnel**, they **can reach the Sun's surface and be emitted as cosmic rays**.

TUNNELING MAGNETORESISTANCE

Computer **hard disks** and **USB sticks** depend on technology called **tunneling magnetoresistance**. **Memory** in a device is stored **using electronic charge**, which is **cleared (deleted) by electrons tunneling through a barrier**.

QUANTUM TUNNEL? = MAGIC?

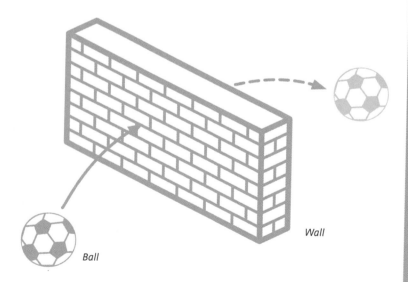

TOUCH SCREENS

The touch screens in smart devices contain **nanoparticles embedded in a polymer film**. By applying a **change in pressure**, such as when pressing on the film with your finger, the rate of tunneling **across the polymer barrier and between the nanoparticles** is **increased by a huge degree**. This drastically **changes the rate at which electrons will** flow via the tunneling effect.

ALGORITHMS AND AI

AIs or "Artificial Intelligence" systems are made from multiple, highly detailed algorithms and have applications in the arts, industry, diagnostics, data analysis, and business. They are far from intelligent, however, and can magnify human biases.

MACHINE LEARNING

Machine learning requires data that feeds information back into algorithms that **fine-tune and improve performance** over time. This has proved beneficial in the **diagnosis and classification of pathologies** in biomedical images. The term "learning" is used when exposing a system to data sets (e.g., new scans) adjusts their operation automatically.

NEURAL NETS

A neural net is a **mathematical model** that **makes decisions based on comparing evidence** (data).

MACHINE BIAS

AIs are marketed as **intelligent entities that save time and money**. Training algorithms with limited data sets leads to biases that become magnified over time, resulting in harmful discriminatory decision-making and actions—e.g., in the context of **calculating insurance premiums, in dating apps, search engines, and in police profiling databases**.

Algorithm: a **series of instructions** written by people; the instructions sometimes operate in loops.
Sorting algorithm: organizes and searches by **looping through arrays of data**—scanning and swapping information to the order required.
Merge sort: this **splits** arrays, **orders** arrays, then **merges** arrays.
Graph search: finds the **fastest route** between two points.
Complexity: the complexity of an algorithm depends on the **number of steps it involves**.
Null character: the **end of a string of values**, indicated with the word (zero) in brackets.
Matrix: an array of arrays with **various dimensions**.
Related variables: can be organized—e.g., as **structs**.
Node: **data points** in a **network**.
Queue: first in, first out.
Stacks: last in, first out.
Trees: the **top node** is called the **root**. **Nodes beneath** are called **children**, **nodes above** these are **parents**. **Leaf nodes** are **at the ends**.

RUBIK'S CUBE ALGORITHMS

To solve the Rubik's cube, follow a series of instructions

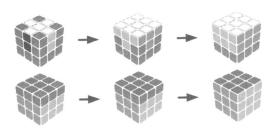

DATA STRUCTURES

- Data must be organized to be accessed.
- Arrays of data can be stored as variables.
- Indexes are used to organize arrays.
- Indexes are indicated within square brackets [0, 1, 2, 3].
- The first number is always index 0.

TECHNOLOGY

INSPIRING SCIENTISTS

Ali Abdelghany (born 1944), marine biologist
Alice Ball (1892–1916), chemist, created treatment for leprosy
Allan Cox (1926–87), geophysicist
Ana María Flores (1952–), engineer
Annie Easley (1933–2011), rocket scientist
Antonia Novello (1944–), physician, US Surgeon General
Bessie Coleman (1892–1921), aviator
Betty Harris (1940–), chemist
Bruce Voeller (1934–94), biologist, AIDS researcher
Burçin Mutlu-Pakdil, astrophysicist
Carl Sagan (1934–96), astrophysicist
Caroline Herschel (1750–1848), discovered comets
Carolyn Porco (1953–), planetary scientist
Catherine Feuillet (1965–), molecular biologist
Claudia Alexander (1959–2015), planetary scientist
Clyde Wahrhaftig (1919–94), geologist and environmentalist
Edith Farkas (1921–93), measured ozone
Eileen McCracken (1920–88), botanist
Eleanor Josephine Macdonald (1906–2007), cancer epidemiologist
Elsa G. Vilmundardóttir (1932–2008), geologist
Eva Jablonka (1952–), biologist, philosopher
Flemmie Pansy Kittrell (1904–80), nutritionist
Fumiko Yonezawa (1938–), theoretical physicist
Gloria Lim (1930–), mycologist
Grace Oladunni Taylor (1937–), chemist
Har Gobind Khorana (1922–2011), biochemist
Haruko Obokata (1983–), stem cell scientist
Heather Couper (1949–), astronomer, educator
Helen Rodríguez Trías (1929–2001), pediatrician
Idelisa Bonnelly (1931–), marine biologist
Jane Wright (1919–2013), oncologist
Jeanne Spurlock (1921–99), psychiatrist
Jeanne Villepreux-Power (1794–1871), marine biologist
Jeannette Wing (1956–), computer scientist
Jewel Plummer Cobb (1924–2017), biologist
John Dalton (1766–1844), relative atomic weights
Kalpana Chawla (1961–2003), astronaut
Katherine Bouman (1989–), computer scientist
Kono Yasui (1880–1971), cytologist
Krista Kostial-Šimonović (1923–2018), physiologist
Lene Hau (1959–), slowed light, briefly stopped a photon
Linda B. Buck (1947–), olfactory receptors
Lydia Villa-Komaroff (1947–), cellular biologist
Mamie Phipps Clark (1917–83), social psychologist
Maria Abbracchio (1956–), pharmacologist, purinergic receptors
Maria Tereza Jorge Pádua (1943–), ecologist
Marianne Simmel (1923–2010), psychologist, phantom limbs
Marianne V. Moore (graduated 1975), aquatic ecologist
Marie M. Daly (1921–2003), chemist
Martha E. Bernal (1931–2001), psychologist
Maryam Mirzakhani (1977–2017), mathematician, Fields Medal
Meghnad Saha (1893–1956), chemical and physical conditions on stars
Melissa Franklin (1957), particle physicist
Michiyo Tsujimura (1888–1969), agricultural biochemist
Mileva Marić (1875–1948), physicist
Mina Bissell (1940–), oncologist
Neil Divine (1939–94), astrophysicist
Niels Bohr (1885–1962), alpha particles, atomic structure
Nora Volkow (1956–), psychiatrist
Patricia Suzanne Cowings (1948–), psychologist
Priyamvada Natarajan (graduated 1993), astrophysicist
Ragnhild Sundby (1922–2006), zoologist
Rohini Godbole (1952–), physicist
Rosalyn Sussman Yalow (1921–2011), medical physicist
Rosemary Askin (1949–), Antarctic research
Ruth Winifred Howard (1900–97), psychologist
S. Josephine Baker (1873–1945), first child hygiene department in New York City
Sally Ride (1951–2012), astronaut, physicist
Sarah Stewart (1905–76), microbiologist
Satyendra Nath Bose (1894–1974), quantum theory
Sau Lan Wu (graduated 1963), particle physicist
Seetha Coleman-Kammula (1950–), chemist, plastics designer
Shirley Jackson (1916–65), nuclear physics
Sonia Alconini (1965–), archeologist, Lake Titicaca basin
Sonja Kovalevsky (1850–91), mathematician
Sophia Getzowa (1872–1946), pathologist
Stephanie Kwolek (1923–2014), chemist, inventor of Kevlar
Stephen Jay Gould (1941–2002), paleontologist
Tanya Atwater (1942–), geophysicist, marine geologist
Toshiko Yuasa (1909–80), nuclear physicist
Una Ryan (1941–), heart disease, biotech vaccines
Valerie Thomas (1943–), invented the Illusion Transmitter
Vandika Ervandovna Avetisyan (1928–), botanist, mycologist
Velma Scantlebury (1955–), transplant surgeon
Vera Danchakoff (1879–1950), cell biologist, embryologist
Xide Xie (Hsi-teh Hsieh) (1921–2000), physicist
Zhenan Bao (1970–), chemical engineer and materials scientist

DATA SHEETS

PREFIX

Prefixes make it easier to express very large or small numbers. It is easier to write 1 mm than 0.001 cm. Prefixes also make it easier to do calculations.

Prefix	Symbol	Meaning	Decimal Place
yotta-	Y	10^{24}	1,000,000,000,000,000,000,000,000
zetta-	Z	10^{21}	1,000,000,000,000,000,000,000
exa-	E	10^{18}	1,000,000,000,000,000,000
peta-	P	10^{15}	1,000,000,000,000,00
tera-	T	10^{12}	1,000,000,000,000
giga-	G	10^{9}	1,000,000,000
mega-	M	10^{6}	1,000,000
kilo-	k	10^{3}	1,000
deci-	d	10^{-1}	0.1
centi-	c	10^{-2}	0.01
milli-	m	10^{-3}	0.001
micro-	μ	10^{-6}	0.000,001
nano-	n	10^{-9}	0.000,000,001
pico-	p	10^{-12}	0,000,000,000,001
femto-	f	10^{-15}	0.000,000,000,000,001
zepto-	z	10^{-21}	0.000,000,000,000,000,000,001

BASE UNITS OF THE SI SYSTEM

Base units are not combinations of other units.

Unit name	Unit symbol	Quantity name	Quantity symbol	Dimension symbol
meter	m	length	l, x, r	L
kilogram	kg	mass	m	M
second	s	time	t	T
ampere	A	electric current	I	I
kelvin	K	thermodynamic temperature	T	Θ
candela	cd	luminous intensity	I_v	J
mole	mol	amount of substance	n	N

DERIVED UNITS AND QUANTITIES

Derived units are combinations of base units. The SI unit of force is kg.m/s² and is also known as a newton. Derived units such as the newton depend on other units.

Quantity	Units / derived units	Expression in terms of SI units
area	square meter	m^2
volume	cubic meter	m^3
speed or velocity	meter per second	m/s (the same as ms^{-1})
acceleration	meter per second squared	m/s^2 or ms^{-2}
wave number	reciprocal meter	$1/m$ or m^{-1}
mass density	kilogram per cubic meter	kg/m^3 or kgm^{-3}
newton per meter	newton per meter N/m or joule per square meter J/m²	$kg \cdot s^{-2}$
specific volume	cubic meter per kilogram	m^3/kg or $m^3\,kg^{-1}$
current density	ampere per square meter	A/m^2 or Am^{-2}
magnetic field strength	ampere per meter	A/m or Am^{-1}
amount-of-substance concentration	mole per cubic meter	mol/m^3 or $mol\,m^{-3}$
luminance	candela per square meter	cd/m^2 or $cd\,m^{-2}$
expended energy	joule second	$m^2 \cdot kg \cdot s^{-1}$
specific energy	joule per kilogram	$m^2 \cdot s^{-2}$
pressure	joule per cubic meter	$m^{-1} \cdot kg \cdot s^{-2}$

QUADRATIC EQUATIONS

Quadratic equation in Standard Form: $ax^2 + bx + c = 0$
Quadratic formula: $x = \left(-b \pm \sqrt{b^2 - 4ac}\right)/2a$

GEOMETRICAL EQUATIONS V

arc length = $r\theta$
circumference of circle = $2\pi r$
area of circle = πr^2
curved surface area of cylinder = $2\pi rh$
volume of sphere = $4\pi r^3/3$
surface area of sphere = $4\pi r^2$
Pythagoras's theorem: $a^2 = b^2 + c^2$

CIRCULAR MOTION

magnitude of angular speed: $\omega = v/r$
centripetal acceleration: $a = v^2/r = \omega^2 r$
centripetal force: $F = mv^2/r = m\omega^2 r$

WAVES AND SIMPLE HARMONIC MOTION

wave speed: $c = f\lambda$
period (frequency): $f = 1/T$
diffraction grating: $d \sin\theta = n\lambda$
acceleration: $a = -\omega^2 x$
displacement: $x = A \cos(\omega t)$
speed: $v = \pm \omega \sqrt{(A^2 - x^2)}$
maximum speed: $v_{max} = \omega A$
maximum acceleration: $a_{max} = \omega^2 A$

ASTRONOMICAL DATA

Sun mass kg.: 1.99×10^{30}
Sun radius m.: 6.96×10^8
Earth mass kg.: 5.97×10^{24}
Earth radius m.: 6.37×10^6
One astronomical unit = Earth–Sun distance = 1.50×10^{11} m.
Light year = distance traveled by light in 1 year
= 5,878,499,810,000 miles (nearly 6 trillion miles)
= 9,460,000,000,000 kilometers
= 9.46×10^{15} m.

CONSTANTS OF NATURE

Quantity	Symbol	Value	Units
speed of light in vacuum	c	3.00×10^8	$m\,s^{-1}$
Planck constant	h	6.63×10^{-34}	$J\,s$
reduced Planck constant (Dirac constant)	$\hbar$ (h-bar)	$1.05457182 \times 10^{-34}$	$J\,s$
Avogadro constant	N_A	6.02×10^{23}	mol^{-1}
permeability of free space	μ_o	$4\pi \times 10^{-7}$	$H\,m^{-1}$
permittivity of free space	ε_o	8.85×10^{-12}	$F\,m^{-1}$
magnitude of the charge of electron	e	1.60×10^{-19}	C
gravitational constant	G	6.67×10^{-11}	$N\,m^2\,kg^{-2}$
molar gas constant	R	8.31	$J\,K^{-1}\,mol^{-1}$
Boltzmann constant	k	1.38×10^{-23}	$J\,K^{-1}$
Stefan constant	σ	5.67×10^{-8}	$W\,m^{-2}\,K^{-4}$
Wien constant	α	2.90×10^{-3}	$m\,K$
electron rest mass (equivalent to 5.5×10^{-4} u)	m_e	9.11×10^{-31}	kg
electron charge/mass ratio	e/m_e	1.76×10^{11}	$C\,kg^{-1}$
proton rest mass (equivalent to 1.00728 u)	m_p	1.673×10^{-27}	kg
proton charge/mass ratio	e/m_p	9.58×10^7	$C\,kg^{-1}$
neutron rest mass (equivalent to 1.00867 u)	m_n	1.675×10^{-27}	kg
alpha particle rest mass	$m\alpha$	6.646×10^{-27}	kg
gravitational field strength	g	9.81	$N\,kg^{-1}$
acceleration due to gravity	g	9.81	$m\,s^{-2}$
atomic mass unit (1u is equivalent to 931.5 MeV)	u	1.661×10^{-27}	kg

GRAVITATIONAL FIELDS
force between two masses: $F = Gm^1m^2/r^2$
gravitational field strength: $g = F/m$
magnitude of gravitational field strength in a radial field:
$g = GM/r^2$

ELECTRIC FIELDS
force between two point charges: $F = (1/4\pi\varepsilon o) \times (Q1Q2/r^2)$
force on a charge: $F = EQ$
field strength for a uniform field: $E = V/d$

THERMAL PHYSICS
energy to change temperature: $Q = mc\Delta\theta$
energy to change state: $Q = ml$
gas law: $pV = nRT$
$pV = NkT$
kinetic theory model $pV = 1/3\,Nm\,(crms)^2$

GLOSSARY

abiogenesis: (noun) the emergence of life forms from nonliving chemical systems

accuracy: how well a measurement approximates to the correct value of a measurement

activation energy: (noun) the energy required to initiate a chemical reaction or process; abbreviated EA

anomaly: (noun) a deviation from expected values

basalt: (noun) dark, fine-grained igneous rock formed from iron- and magnesium-rich lava; the main component of the oceanic crust and lunar regolith

beam: (noun) a ray or shaft of light from a source

binary stars: (noun) a pair of stars that orbit a common mass

blueshifted: when an astrological object is observed to move toward Earth

bond length: (noun) distance and most stable position between atoms in a bond

Boyle's law: (noun) for a fixed amount of gas at a stable temperature, the gas's volume is inversely proportional to its pressure

buffer: (noun) substance that compensates for concentration of hydrogen ions (H^+) and maintains a fairly constant pH

calibrate: (verb) to check and correct the accuracy of an instrument

carbon dating: uses the naturally occurring isotope carbon-14 to determine the age of carbonaceous (carbon-containing) substances

Cartesian plane: (noun) a rectangular coordinate system that uses (x, y), where the x value is the horizontal coordinate and the y value is the vertical

chromatin: (noun) protects DNA in the nucleus; only found in eukaryotes

cohesion: (noun) interaction between molecules in a substance

confirmation bias: (noun) when information/results are interpreted as confirming one's preconceptions and information/interpretations that contradict these beliefs are ignored

constant: (noun) a quantity with a fixed value

coulomb: (noun) a metric unit of electrical charge = 6.24×10^{18} electrons

cytosol: (noun) the fluid of cytoplasm where metabolism takes place; made of water and fibrous proteins

De Broglie wavelength: (noun) $\lambda = h/p$ where wavelength (λ) of a fundamental particle is related to momentum (p) and Planck's constant (h)

dependent variable: (noun) a variable (changeable) parameter in an experiment or observation

deuterium: (noun) a stable isotope of hydrogen with a neutron in its nucleus

ENIAC: (acronym) electronic numerical integrator and computer; first general-purpose electronic computer

evidence: (noun) support for an opinion or hypothesis

exothermic: (adjective) a reaction that releases heat

fluorescence: (noun) an emission of light following the absorption of light or energy by a substance; related to atomic emission and line spectra

fossil: (noun) the preserved impression/remains of an organism, its tissues replaced by minerals

genetic drift: (noun) random changes in gene frequencies

ground state: (noun) lowest energy state—electrons fill the lowest energy levels first

Hardy–Weinberg equilibrium: (noun) the frequencies of alleles and genotypes remain constant from generation to generation in the absence of evolutionary forces

ice core: (noun) a cylinder of ice containing layers of snow that have compacted into ice over long periods of time

ideal gas: (noun) a theoretical gas with no volume or intermolecular forces of attraction or repulsion; an ideal gas equation uses gas constant, r

independent variable: (noun) a parameter changed in a controlled way during an experiment

Intergovernmental Panel on Climate Change (IPCC): (organization) a group of international scientists, established in 1988, who evaluate the risk of human-caused climate change

kinetic molecular theory: (noun) theoretical description of molecules and kinetic energy; energy of motion = temperature

LANL: (acronym) Los Alamos National Laboratory

lichen: (noun) fungi that live symbiotically with photosynthetic algae or bacteria

limiting factor: (noun) any essential resource an organism needs for survival that is in short supply in its environment

line spectra: (noun) emission spectrum with sharply defined lines; lines correspond to a wavelength of light given off during an electron transition from an excited state to the ground state

lunar eclipse: (noun) the Earth is aligned between the Sun and Moon

qualitative: verbal description of observations or data, i.e., nonnumerical/not numerically measured

quantitative: measurements and observations expressed numerically

radian: SI unit of plane angle; there are 2ϖ radians in a circle

radiation belt: regions of charged particles in a magnetosphere

radical: atomic or molecular species with unpaired electrons

sapping: erosion where water leaks upward through porous rocks

Scleractinian corals: hard reef-building corals

Sea of Tranquillity: lunar landing site of Apollo 11 on July 20, 1969

secondary air pollutants: pollutant formed when primary pollutants react, e.g., acid rain: sulfur dioxide or nitrogen oxides reacting with rainwater

tar sands: sandy deposits containing bitumen with a high sulfur content

ubiquitin: protein found in the cytoplasm

ultraluminous galaxies: galaxies bright with infrared wavelengths

umbra: the dark central region of a sunspot

upwelling: wind-driven rising of dense, cool, nutrient-rich water to the ocean surface

vacuum: a space completely devoid of any matter

vector: has magnitude, length, and direction

vermicomposting: earthworms create compost and enhance the process of waste conversion

viability: an organism's chances of completing its life cycle: surviving to maturity

x-band: radio frequencies extending between 5,200 and 10,900 MHz

yellow fever: an acute viral disease spread by the *Aedes aegypti* mosquito

zodiac: twelve constellations dividing the ecliptic into approximately equal parts

zooplankton: microscopic animals that live in water

zooxanthellae: single-celled yellow-brown algae (dinoflagellate) in symbiosis with reef-building coral

zygote: a fertilized egg containing two sets of chromosomes

FURTHER READING

Roma Agrawal, *Built: The Hidden Stories Behind Our Structures*

George Basalla, *The Evolution of Technology*

Gurminder K. Bhambra, Dalia Gebrial, et al., *Decolonising the University*

Rachel Carson, *Silent Spring*

T. J. Demos, Basia Irland, et al., *Elemental: 1: An Arts and Ecology Reader*

Richard Feynman, *Six Not So Easy Pieces*

Sam Kean, *The Disappearing Spoon: And Other True Tales of Madness, Love, and the History of the World from the Periodic Table of the Elements*

Mark Miodownik, *Liquid: The Delightful and Dangerous Substances That Flow Through Our Lives*

Safiya Noble, *Algorithms of Oppression*

Cathy O'Neil, *Weapons of Math Destruction*

Roger Penrose, *Road to Reality*

Gina Rippon, *The Gendered Brain: The New Neuroscience That Shatters the Myth of the Female Brain*

Angela Saini, *Inferior*

Angela Saini, *Superior*

Anna Tsing, *The Mushroom at the End of the World*

Anna Tsing, Heather Anne Swanson, Elaine Gan, and Nils Bubandt (eds), *Arts of Living on a Damaged Planet: Ghosts and Monsters of the Anthropocene*